IMechE
Seminar Publication

I MECH E

Noise and Vibration
Advances in Research and Development

Organized by
The Research and Technology Committee of
The Institution of Mechanical Engineers (IMechE)

Sponsored by
LMS International and co-sponsored by the Institute of Physics

IMechE Seminar Publication 2000–3

**Professional
Engineering
Publishing**

Published by Professional Engineering Publishing Limited for The Institution of Mechanical Engineers, Bury St Edmunds and London, UK.

First Published 2000

ISSN 1357–9193
ISBN 1 86058 277 X

A CIP catalogue record for this book is available from thc British Library.

Printed by Antony Rowe Limited, Chippenham, Wilts, UK.

Contents

S682/001/99 **Vibro-acoustic modelling – where we are, where we should go, and how?**
C F McCulloch, L Cremers, and P Guisset 1

S682/002/99 **Visions and realities in vehicle noise and vibration**
P van der Linden, K Wyckaert, and C F McCulloch 15

S682/003/99 **Understanding and controlling acoustic effects of outdoor surfaces**
K Attenborough 29

S682/004/99 **Sound power flux measurements in exhaust systems**
P O A L Davies, K R Holland, and D van der Walt 39

S682/005/99 **Theoretical prediction of ground vibrations generated by road vehicles**
V Krylov 51

S682/006/99 **European co-ordinated railway noise research**
B Hemsworth 69

S682/007/99 **A review of computer modelling of sound in buildings**
B M Shield, S Dance, and L N Yang 77

S682/008/99 **Aircraft noise**
Technical Note A J Kempton 85

Authors' Index 91

Related Titles of Interest

Title	Editor/Author	ISBN
IMechE Engineers' Data Book	Clifford Matthews	1 86058 175 7
Vehicle Noise and Vibration	IMechE Conference Transactions	1 86058 145 5
Vibrations in Rotating Machinery	IMechE Conference Transactions	1 86058 009 2
Automotive Modelling and NVH – Techniques and Solutions	IMechE Seminar Publication	1 86058 124 2
Noise in Fluid Machinery	IMechE Seminar Publication	1 86058 246 X
Tomorrow's World	IMechE Seminar Publication	1 86058 187 0

For the full range of titles published by Professional Engineering Publishing contact:

Sales Department
Professional Engineering Publishing Limited
Northgate Avenue
Bury St Edmunds
Suffolk
IP32 6BW
UK

Tel: +44 (0)1284 724384
Fax: +44 (0)1284 718692

S682/001/99

Vibro-acoustic modelling – where we are, where we should go, and how?

C F McCULLOCH, L CREMERS, and **P GUISSET**
LMS International, Leuven, Belgium

1 SYNOPSIS

The paper presents an overview of methods for computer modelling to predict the acoustic and vibro-acoustic behaviour of interior and exterior fluid regions and their interaction with structures: simplified and semi-empirical approaches, geometrical and energy-based methods (acoustic ray-tracing, statistical energy analysis, ...) and wave-based methods (structural finite elements, acoustic finite elements, boundary elements, infinite elements, ...).

Based on the modelling procedures and limitations of these methods, their usefulness in different applications in mechanical engineering is discussed, with particular reference to vehicles (automotive, rail, aerospace) other applications in underwater fluid-structure problems, and noise from machinery, structures and consumer products, with examples.

The barriers to wider use of these methods, with particular focus on the technical limitations of modelling capabilities, is discussed, and the needs for further developments are outlined, with some possibilities for future techniques including improved speed and ease-of-use, multi-physics behaviour, such as noise with fluid flow and flow-induced noise, exterior vehicle pass-by noise prediction, multiple interacting coherent and incoherent sources and paths, auralization and sound quality metrics and prediction. Some current developments are described, and various issues are suggested on which partnerships of academic researchers, software developers and industrial users can focus, to develop practical improvements to current methods.

2 INTRODUCTION AND BACKGROUND

2.1 Noise and vibration – a design issue

Noise is becoming an increasingly-important phenomenon: it is an issue in the lives of many people in urban societies and even in rural areas. Noise from engineered products such as vehicles, aircraft, power plant and machinery is unwanted, and in the many markets also provides a subjective indicator of the quality of such products. Vibration can also be an issue in itself, from a comfort or health aspect, or more generally it is at issue as the source of noise.

In order to minimise cost, it is desirable to eliminate, reduce, or in some way control, noise and vibration at the design stage of engineered products, and to implement these controls at minimum cost on the product. Thus, predictive methods are increasingly attractive, to provide more, better, more-accurate, information about noise, in a timely manner: early in the design process, before design change becomes expensive; and later in the design process, where test-based design optimisation and refinement can be costly.

This paper sets out to review the predictive methods for noise and vibration and what may be needed to make them more effective in future in working towards fulfilling these desires.

2.2 The computing environment: hardware trends and ease of use

Computer-based modelling methods are by definition dependent on the computer hardware and environment within which they operate. In considering where we are today in computer modelling of noise and vibration, it is worth remembering that the hardware environment has changed radically even within the last decade, during which computational acoustics has become common. Computer manufacturers have talked of a price-performance increase of the order of $x10^3$ per decade, and this is not far from reality: compare the modern desk-top PC (or even portable) with a 'workstation' or office-sized machine of 1990. With more powerful computers, 'local' machines are now capable of what was considered super-computing only a short while ago. These trends will undoubtedly continue, and will allow increasingly-large and complex problems to be solved numerically without excessive time or resource requirements. Likewise, more rapid solutions will allow more alternatives to be considered, a pre-requisite for optimisation processes.

In parallel with the computer power available, the software operating environment has changed radically, from one in which data are prepared and submitted almost 'off-line' to one in which most modelling work is done interactively, with a lot of graphical feedback, and with scanning and visualisation tools, to aid the interpretation of results. This makes it much easier to create, solve and assess a model, although we should note the danger of getting easy access to the *wrong* results, from conceptual or data errors!

3 CURRENT ACOUSTIC MODELLING: A REVIEW

3.1 The physics of sound and vibration

Sound is essentially a wave phenomenon of compressions and rarefactions in the fluid. (Here, we will not consider more complex acoustic waves in solids or viscous fluids). Vibrations in structures can consist of combinations of compressive waves and (more often) bending and shear. The interaction of a vibrating structure and an adjoining fluid, with kinematic and material continuity considerations at the interface, defines vibration-generated noise radiation or noise-induced vibration, in principle always with a two-way interaction, but in practice the 'feedback' element may be so small as to be negligible.

At higher frequencies, or if there are many combinations of waves interacting with each other, for example with a high modal density, it may be reasonable to assume that the sound or vibration are behaving more as 'energy' which is stored or propagated through the structure and the fluid. Whether this behaviour is derived from a detailed analysis of the wave behaviour, summed in a 'random' or incoherent manner, or is used as a basic assumption in a simpler modelling approach, is a key decision for the modeller.

3.2 Acoustic and vibro-acoustic modelling approaches

A simple division of acoustic modelling methodologies can be made, into different approaches based on the assumptions about the acoustic (and vibro-acoustic) behaviour:

- empirical methods, based on simple 'rules' or formulae, often derived from mcasurements; these may even be amenable to hand calculation in simple cases
- special analytical methods, usually only relevant to particular cases (transmission loss in ducts, one-dimensional propagation, flow noise sources...)
- general, three-dimensional, sound dispersion and structural energy flow models (ray- or beam-tracing, statistical energy analysis, ...)
- general, three-dimensional, wave-based and multi-physics models (finite elements, boundary elements, infinite elements, ...).

The last two approaches are the more interesting, because they offer generalised methods that can be adapted and used for many different applications, as will be discussed below. They are also embodied in general-purpose, commercial, supported, software.

4 VIBRO-ACOUSTIC MODELLING TOOLS

4.1 Geometrical acoustics

Geometrical acoustics uses a ray-tracing method to calculate the way sound is distributed. (More precisely, beam-tracing is used, which gives more robust results). The method can be used for completely-closed interior spaces, partially-open ones and fully-open exterior environments. Each source is located in space and given a defined power. It may radiate equally in all directions, preferentially in some (by a defined directivity, *eg* a directional loudspeaker) or wholly within a defined region (*eg* a vibrating panel, radiating on one side only). Each source may also be basically coherent or incoherent. The rays (beams) from each source are traced around the defined space and reflect off all the surfaces they contact like mirrors (see Figure 1) losing energy at each reflection in accordance with the absorption properties of the surfaccs. The rays 'capture' receiver points (occupants head location, ...) as their cross-sections pass by, defining an echo with a certain arrival time (due to its path length from the original source) and a certain energy level (due to the energy absorbed at the reflections and the spherical spreading-out of the beam cross-section).

From the echograms (impulse responses) at each receiver point (Figure 2) many sound parameters can be derived: steady-state SPL, acoustic quality measures and Speech Transmission Index. All of these can be frequency-dependent, since both absorption properties and source powers and directivities can vary with frequency. Figure 3 shows an example of a noise map in a vehicle cabin, due to certain equipment noise. By superimposing many different sources, varying their properties, and so on, with intelligent data-handling in the modelling process, many different possible scenarios can be assessed and the effects of different combinations of sources (wanted or unwanted) can be rapidly analyzed. Typically, the basic model data such as geometry are loaded more-or-less automatically from a general-purpose CAD system, but the more-specialized acoustic data such as loudspeaker properties are usually derived elsewhere and saved in dedicated databases.

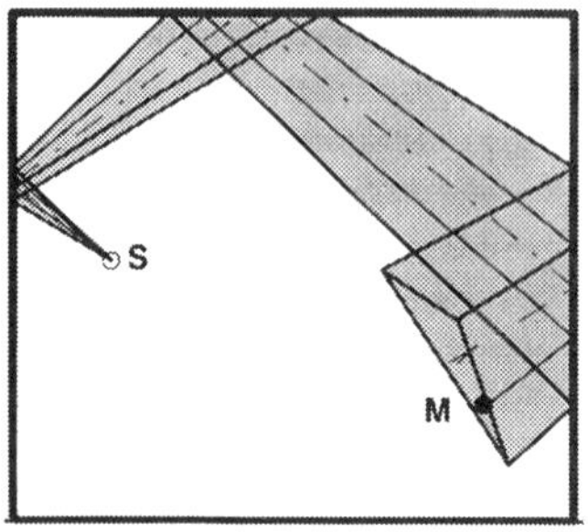

Figure 1: Ray- (beam-) tracing scheme

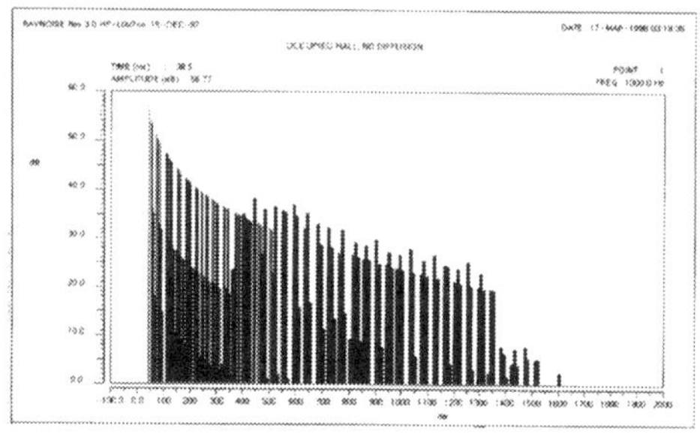

Figure 2: Echogram from beam-tracing

Special techniques exist for modelling diffusion, diffraction (*eg* over barriers) and the transmission of sound through partitions from one region to another (*eg* from outside to inside of a vehicle, from one compartment to another, ...). These techniques enable a complete model of the noise environment around as well as inside a vehicle or building to be built up.

4.2 Statistical Energy Analysis

Geometrical acoustics is more-or-less limited to purely acoustic calculations, such as building and environmental acoustics, and noise distribution in or around a vehicle. Where the noise is mainly structure-borne, it is necessary to consider mainly the structural vibration behaviour, rather than the geometrical distribution of the sound. At medium to high frequencies (to be discussed later) the standard approach for modelling has become Statistical Energy Analysis (SEA). In this approach, the structure is broken-down into sub-systems, each of which has defined boundaries, geometry (curved shell, acoustic space, ...) properties (thickness, honeycomb core, beam cross-section, ...) and connections to other sub-systems (point, line, area, ...). The geometry, properties and connections define the vibrational characteristics of the sub-systems and their interactions. An important input is also information about damping ('loss factors') within the sub-systems and in their connections. The model, consisting of the assembly of all the sub-systems, is driven by energy inputs related to the physical sources of vibration and noise, such as suspension mounts, motors, ancillary equipment, and so on, and an energy-balance (or flow) is derived. The results are then the energy levels (vibrational or acoustic) in the sub-systems, and the flow of energy from the sources into the dissipating systems.

SEA provides an effective modelling tool for the global behaviour of structural vibrations and noise radiation to the exterior and into interior spaces, with the benefit of being reasonably fast and therefore practical at the design stage. By integrating experimental vibration measurements on prototypes, previous designs, or sub-systems of a new design, the data input to the models can be validated and improved. Whilst an SEA model is essentially just a network of sub-systems and connections, it can also be visualised in a geometrical way.

4.3 Finite Element and Boundary Element Methods

Finite element (FE) methods are well-established for structural models, including dynamics as well as stress analysis. FE acoustics models are effective, where discrete wave behaviour is to be modelled. Since the wave behaviour is 'sampled' in space, a criterion for minimum element size versus wavelength is applied, typically six linear elements per wavelength. By definition, FEM models are closed (finite) so cavity modes can be extracted, and forced response to structural vibration or other boundary conditions can be solved using modal superposition as well as direct methods. A FEM equation system is characteristically sparse, so is also amenable to some efficient solvers. Because each element has independent matrices, which build up the system matrices, each element can have unique properties: sound speed and density (*eg*, changing due to temperature) and even volumic absorption.

Boundary element (BE) acoustics models implicitly have the Somerfeld infinite-field radiation characteristic build-in, if they are exterior models, unlike FEM which is conventionally limited to interiors. BEM can also be used for interior closed geometries, and the most-useful BEM formulation (Indirect, Variational method) is generalized and can be used for interiors or exteriors or 'mixed' problems (cavities with openings...) and can handle thin appendages like ribs, and complex topologies with 'junctions' (T- or X-intersections of surfaces). Half-space conditions or symmetry can be used.

Both FE and BE acoustics models can be coupled to FEM structural models in an effective way, for full, two-way, fluid-structure interaction. A 'modal coordinates' description of the structure is efficient in this context. Boundary conditions can then include: known structural vibrations; known forces on the structure; and known incident wave fields. Most of the modelling done with FE and BE acoustics is currently in the frequency domain (*ie*, Helmholtz form of the wave equation) but time domain methods are also possible. For BEM, time domain solution has not proved reliable, but for FEM it is robust and not too sensitive to input data or solution control parameters such as time step size.

Infinite elements (IFEM) provide an extension of finite elements into exterior, free field, modelling. Both radiation and scattering (incident wave) problems can be handled. The infinite elements have terms representing the wave propagation, built into their element functions: geometrically, they are seen as infinitely-long elements, which begin on bases which are the outside faces of (volumic) finite elements, which are used to model the near field. The base surface usually has to be a regular geometric shape (sphere, ellipsoid...). The benefits of FEM can be used in the near field, and IFEM offers robust time-domain solutions as well as frequency-domain.

4.4 Vibro-acoustic panel and trim modelling

Special vibro-acoustic models using FE methods can be applied to the design of panels and trim assemblies. These models use both structural and fluid elements for the appropriate parts, and special elements with properties related to both fluid (air) and structural 'phases' in the material. Typically, these elements use Biot's method to define the interaction of acoustic waves and structural vibrations, in porous or fibrous material. Thus one can model the behaviour of a structural panel, where the vibrations are altered and the acoustic radiation or transmission is changed by the addition of multiple layers of damping and absorbent material (see Figure 3). This can be especially useful in applications such as design for minimum acoustic transmission of floors, other vehicle panels or doors.

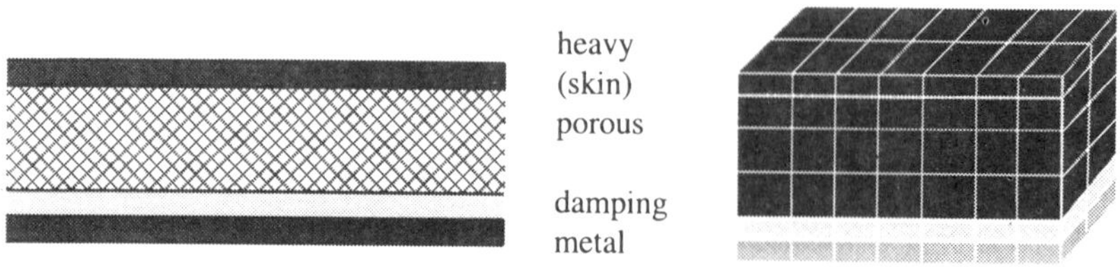

Figure 3: Multi-layer material (floor, door, ...) using poro-visco-elastic vibro-acoustic FE model

5 VIBRO-ACOUSTIC MODELLING APPLICATIONS

5.1 Automotive

Vehicle interiors can be modelled with Finite Elements or Boundary Elements for lower frequencies (Figure 4). The FEM approach offers the possibilities of direct extraction of modal frequencies, although these may well be inaccurate if body-cavity interaction is not included, and modelling of volumic absorbent materials (*eg*, seats). The BEM approach requires a mesh which is usually much easier to produce, so it is more commonly used: fluid-structure interaction is also included, by using a reduced modal model of the structure.

For higher frequencies, energy transmission through the structure is important and SEA can be used, with acoustic regions modelled as distinct sub-systems.Ray-tracing models are used to analyse the behaviour of the in-car audio system.

Exterior noise from powertrain units (engine block, sump, covers, ...) can be predicted with BEM models, through most of the frequency range of interest (Figure 5) or simplified Raleigh integral approaches can be used, with reasonable approximate results, in the upper part of the frequency range. Either the complete powertrain unit can be modelled or component parts can be assessed individually: the latter produces smaller models, easier to process. Other related components can also be modelled: intake systems with BEM (typically, coupled to a structural model, because it is light and relatively flexible and re-radiates noise) and exhaust systems, usually with FEM (in full 3-d, with variations in fluid acoustic properties, and possibly volumic absorbers, impedance through perforated pipes, and other special features, including the influence of flow).

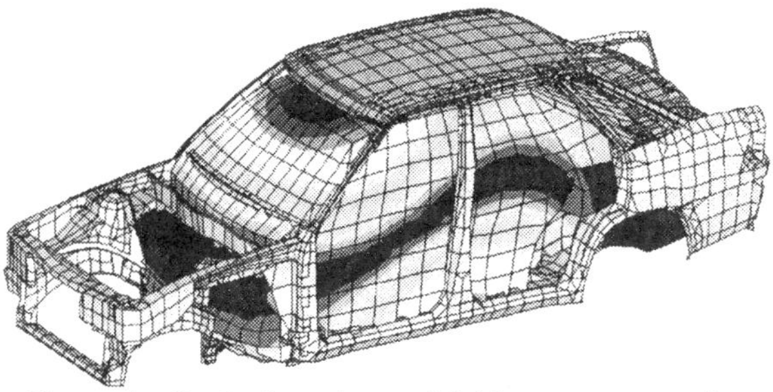

Figure 4: Car body-noise model (closures removed)

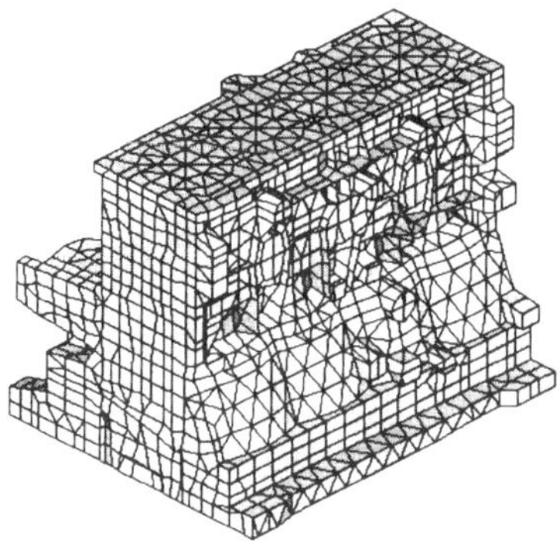

Figure 5: Engine noise radiation (BEM)

Some other exterior noise models are in use, for example finite elements and infinite elements for tyre noise (Figure 6). However, a model for the complete exterior noise radiation of a whole road vehicle, taking into account all mechanisms of noise generation and propagation, does not yet seem to be in existence: this is an area worth some further research.

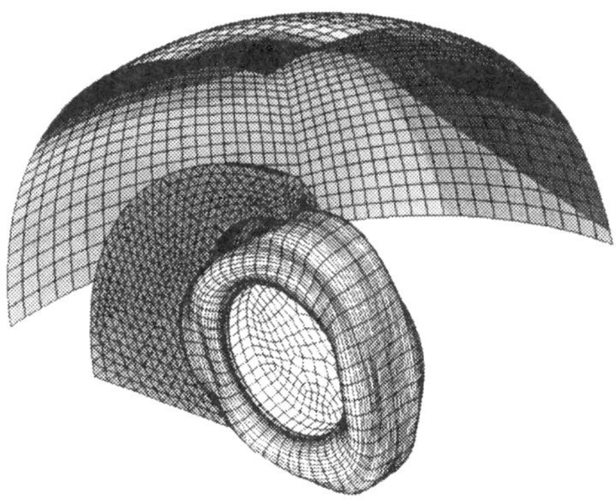

Figure 6: Tyre noise radiation - finite elements and infinite elements

5.2 Railways

Because of the size of rail vehicles, their interior noise is not generally amenable to FEM or BEM acoustic models. Typically SEA is used, although ray-tracing may also be useful especially if noise penetration from exterior sources is an issue (*eg*, machinery noise, in tunnels). Ray-tracing is also of use for interior audio system analysis, for PA *etc*, and for ventilation and similar sources (Figure 7).

For the optimisation of individual panels, the multi-layer vibro-acoustic FEM approach can be used.

'Local' models of individual components (engine/motor, gearbox, intake, exhaust, wheel, ...) can be made with BEM, in a similar manner to the applications on automotive vehicles. (Figure 5, 8).

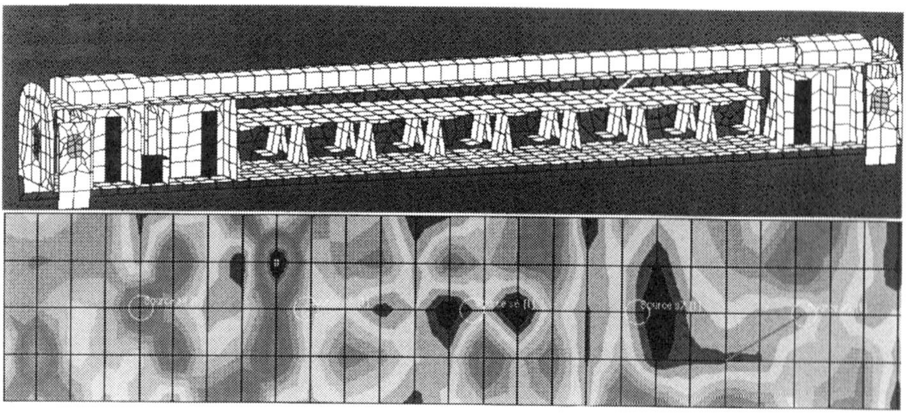

Figure 7: Noise map inside railway vehicle

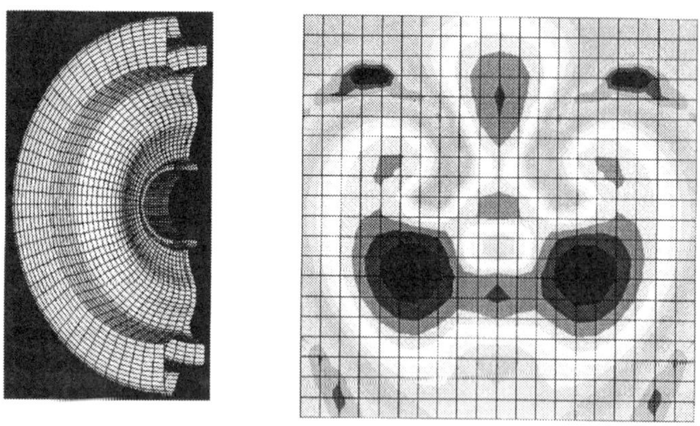

Figure 8: Local model of railway wheel noise radiation (BEM) optimisation study

The exterior noise from the train can be modelled with ray-tracing, both to assess the propagation to the environment, with effects of source height and local or remote shields or barriers (Figure 9) and to assess the reverberant noise that returns onto the vehicle and penetrates to the interior.

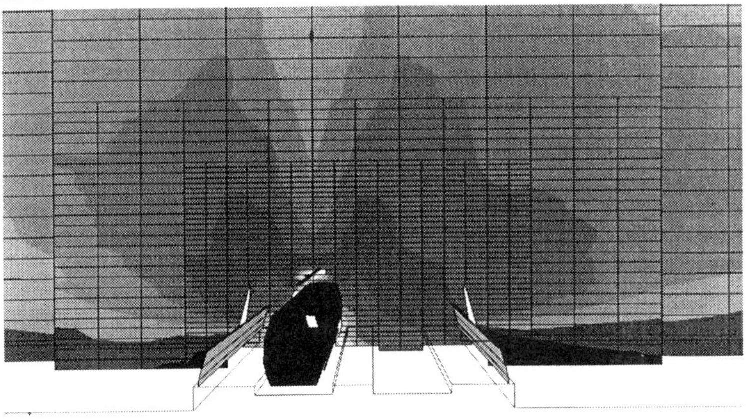

Figure 9: Noise map around high-speed train, with barrier diffraction

For the wheel-rail interaction, a very particular noise and vibration source, a special form of SEA model can be used, to define acoustic and vibration sources for other calculations.

5.3 Aerospace

The interior noise of aircraft can be modelled with BEM, if the external source is localised and especially if it is tonal (typically, a propeller-driven aircraft). A fully-coupled fluid-structure interaction can be performed, with special source terms in the solution to define the rotor noise (detailed discussion is beyond the scope of this paper) and a FEM structural model. The final result is noise distribution in the interior of the structure. If the noise is due to a turbulent flow or a jet, the exterior acoustic pressure loading can be applied directly on the structure, but in this case random terms are required: the exterior noise can be either completely random (*ie*, perfectly diffuse) or can be given some spatial correlation (non-uniform pressure distribution and partial coherence between values at different locations).

On an aircraft, it is often the interior noise that is of interest, plus the radiated propulsion noise. On a spacecraft such as a satellite, the vibration response to a defined acoustic field is often of interest, finally to derive acoustic fatigue response for the launch process. In this case, the acoustic loading will often be a diffuse field, with a given pressure spectrum, and a fully-coupled BEM acoustic analysis is used, with a FEM structural model. The random loading results in spectra (PSDs) of vibration response from which strain and stress PSDs can be derived, as the input to fatigue-damage estimates.

5.4 Sonar and audio devices and other products

In the marine field, fluid-structure interaction is clearly important: hull-water interaction can be modelled with BEM acoustics and FEM structure models, at lower frequencies, for studies of machinery noise radiation and sonar 'signature'. Due to the geometrical size of most vessels (ship, submarine, torpedo, ...) however, at higher frequencies more specialised programs are needed.

The size of sonar devices themselves is suitable for BEM modelling of their radiation characteristics (directivity, beam-forming...) and the interaction of multiple transducers in an array, with FEM models of structural and local acoustic behaviour (mounting windows, ...) fully coupled. (Figure 10).

In air, likewise, loudspeakers can be modelled with fully-coupled BEM-FEM fluid-structure models. (Figure 11). The modelling of the response of the space in which the speakers are placed is better done with ray-tracing, however.

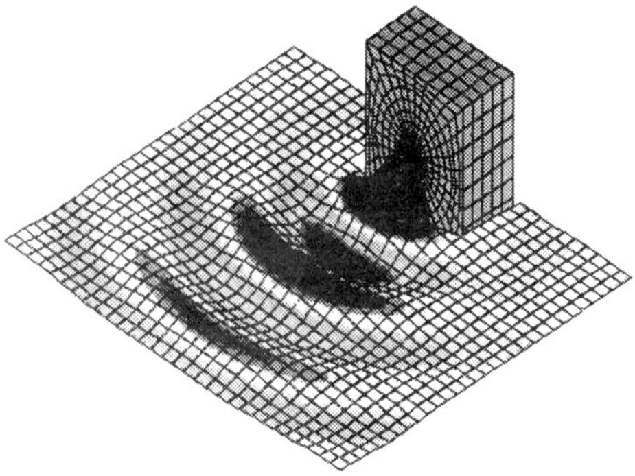

Figure 10: Loudspeaker (coupled BEM)

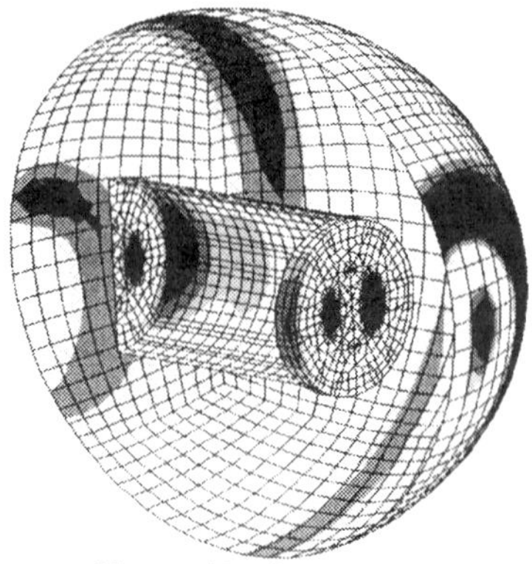

Figure 11: Sonar array with window and field points

Other models of the noise radiation behaviour (wanted or unwanted) of domestic equipment and other products can also be made with BEM acoustics techniques.

5.5 Architecture, construction and environment

As mentioned earlier, ray-tracing (or more precisely, beam-tracing) is effective in modelling 'architectural acoustics': not only the complex results required for fine acoustic design of an auditorium, but also intelligibility of PA systems in public places such as stations and shopping malls, and more basic results for the noise impact on the environment within and around engineering plants. For exterior noise impact assessment, simpler approaches based on propagation formulae (partly empirical in origin) and 'rules' or standards may be adequate.

6 LIMITATIONS IN VIBRO-ACOUSTIC MODELLING

6.1 Frequency range

For the wave-based methods (acoustic BEM, acoustic and structural FEM) it is usual to require a minimum of 6 linear elements per wavelength, for reasonable accuracy, thus setting an upper frequency limit. Conversely, if the upper frequency is raised, the element size decreases and the model size increases, and the calculation time increases as a power of the model size, up to the limit of computer resources (or the time the user is prepared to wait). A limit is also imposed by the assumption of discrete wave behaviour, with distinct modes of the acoustic cavity or the structure, which starts to break down at some frequency limit.

In some applications, such as road vehicles, the problem can arise that the acoustic cavity is still in the region of distinct modes (up to say 200-300 Hz) whereas the structure already has a quite high modal density at a lower frequency (100-200 Hz).

By contrast, the acoustic ray-tracing and SEA approaches have a lower frequency limit, since they make the assumption of high modal density and acoustic or vibrational energy dispersion. In purely acoustic models, the limit can be lowered or even removed, by using phase ray-tracing, in which a coherent wave-like behaviour is added back to the acoustic propagation. It is also suggested that SEA may be used at lower modal densities, but this seems uncertain.

6.2 Linearity

Most of the methods discussed here assume linear behaviour (small perturbations) which is also convenient for solutions in the frequency domain. However, high-energy acoustic behaviour (which may be of interest in acoustic fatigue, for example) becomes non-linear. The structural behaviour may also include non-linearities: contact, non-linear or high-hysteresis materials, ... (for example, in tyre noise models) - or kinematic motion (flexible, multi-body dynamics). Most of these non-linear effects are handled more easily in a time-domain solution: opportunities exist for developments related to various applications.

6.3 Uncertainty

In practical modelling, there is often an uncertainty over the input data for the model: values for acoustic sources or structural vibrations or forces may not be well-quantified, materials data (especially damping) may be uncertain. Ideally, the models should give indications of sensitivities or 'confidence limits' based on the uncertainties of the inputs. In some cases such as SEA (*Statistical* Energy Analysis) this is implicit, but most vibro-acoustic methods are deterministic in nature. Examination of sensitivities then require additional processing.

A key input in a deterministic model such as acoustic BEM, is the phase relationship of different sources or vibrations. This is often difficult to establish accurately, or it may be subject to 'drift' on the real-world product. Some inputs may be incoherent (within themselves, or in relation to others). Handling multiple loadcases, with separate superposition of their results, could assist in addressing this.

6.4 Resources

Although the trend in computer speed and capacity is in the right direction, some of the methods described here require large amounts of computer resources. This is especially true of BEM calculations, where the calculation time *per frequency* can make a complete solution of (for example) whole-engine noise radiation at many frequencies, with several loadings, representing the different design 'run-ups', almost impossible within a reasonable total time. At present, then, the modeller has to use considerable judgement to reduce the numbers of runs to be performed. Real design optimisation and consideration of several alternatives is thus difficult.

There is an opportunity for substantial improvements in the practical usefulness of some of the acoustic modelling methods, by 'smart' approaches to reduce the calculation times.

It must be pointed-out, by contrast, that *some* of the methods described here (*eg*, SEA) have very fast solutions and can be re-run many times without a big impact on total analysis time: in fact, it is ease-of-use and the speed of modelling and re-modelling which is an issue.

7 FUTURE DIRECTIONS

7.1 Ease of use

Making vibro-acoustic modelling easier to use is a key requirement in bridging the gap between its use by specialists in research and development groups, and general product design teams. It is also a requirement to provide more-accessible, interpreted, results. Three areas of development can be suggested:

Integration into the design process, by integrating vibro-acoustic modelling tools into CAD and tying them closely to other CAE tools. Assessing the vibro-acoustic performance of the (virtual) design becomes one concurrent element alongside the other design decisions.

Result interpretation and optimisation, to assist the design team to understand the significance of the vibro-acoustic performance, its sensitivity to design changes and what to do to improve it. Multi-disciplinary optimisation tools provide the opportunity to link together the assessment of many different product performance criteria and other design constraints such as weight and cost. Expert systems, or 'reporting and advisory' tools, are also of interest: the vibro-acoustic modelling becomes part of a holistic design process.

Synthesis of results and hybrid methods, allied to the other two developments, allow the use of the maximum amount of retained, validated, knowledge about product performance. Experimental data and historical information can be included (see: Vanderlinden et al, 'Visions and Realities in Vehicle Noise and Vibration', also in this volume S682, IMechE, October 1999).

7.2 Speed

Despite the rapid increases in computer power, and the efforts of software developers to improve the performance of their codes, calculation time is a significant issue in several applications, especially with acoustic BEM. Faster solutions will allow bigger models and wider frequency ranges, or the consideration of more alternatives, sooner.

New hardware trends (*eg*, parallel processing) must be exploited wherever possible, and new solver algorithms and 'smart' mathematics applied.

7.3 New features and capabilities

Methods for efficient handling of assemblies of components or sources, to consider whole-system noise as well as individual components in detail, should be applied. Tools for the assessment of multiple noise sources and paths are already developed. Integration with other predictive models is of great interest – especially the derivation of forcing functions from multi-body dynamics analyses.

Allied to assembly models is the need to handle the uncertainty and incoherence of the input data for each component, and the consequent uncertainties or variability of their combination.

Modelling of the effects of flow is presently an area of great interest: both the effects of flow on the noise ('wave convection'...) and the generation of noise by flow (turbulence...).

If the flow is steady, the effect on noise propagation is straightforward, although programming this into a general, three-dimensional, model is not. At present, in acoustic BEM for exterior models it has proved unsatisfactory, but in acoustic FEM for interior models it is already possible. The use of infinite elements (IFEM) for the far field, with a near-field FEM mesh, offers the possibility to extend this to general cases for exterior models.

If the flow is unsteady, in a local region such as a turbulent boundary layer or due to vortex shedding, two possibilities are apparent at present: firstly, modelling the whole flow regime include the small perturbations which are the 'acoustic' part, using direct numerical simulation (DNS); secondly, modelling the unsteady flow regime in a 'global' sense with large eddy simulation (LES) and from that deriving results which define local acoustic sources. The latter is a numerical representation of Lighthill's 'acoustic analogy': there is current activity studying this, and opportunities for more research efforts. In theory, DNS is the more correct approach, since it includes all effects in a precise way, but in common with many numerical modelling approaches which attempt a complete physical representation of the behaviour, it suffers from sever practical limitations due to the computational effort required, even to model a small region of a real practical problem.

Most current modelling is in the frequency domain (sometimes using a modal superposition). Solutions in the time domain are of interest, both to process more frequencies over a wider range, efficiently, and to handle transient effects. They can also offer an easier starting-point for extensions to cover non-linear behaviour. BEM solutions in the time domain (retarded potential formulation...) have previously proved unsatisfactory, but FEM and IFEM solutions can be stable and relatively insensitive to input changes or solution controls like time step size.

8 CURRENT AND FUTURE DEVELOPMENTS AND OPPORTUNITIES

8.1 Current development activities

This section highlights some of the on-going research activities with which LMS is associated. Some research activities are internally-driven, or directed by external clients with specific applications in mind, whereas others are carried on the context of EU and other funding frameworks (which still have final applications as an objective) or with university research collaborations. The order of presentation is not significant in terms of time or value.

- Integration of vibro-acoustic BEM and FEM with the CAD system CATIA (ease-of-use).
- Linking BEM acoustic radiation with multi-body dynamics (*eg*, for powertrain noise).
- The use of modal contribution vectors: the acoustic radiation prediction is based on the noise field generated by each mode of vibration of the structure, rather than the operating vibration shape at each frequency; results for individual frequencies or structural excitations are produced by superposition of modal results. The effect is to reduce greatly the solution time for large problems, such as powertrain noise, where many frequencies and loadcases can exist.

- Non-linear acoustic sources: in a FEM acoustic model, wave sources can be defined which have amplitudes dependent on the local acoustic pressure, *ie* with feedback (*eg*, noise generated by combustion in a burner, where formulae for the source-pressure relationships are known).
- The use of Pade expansions for fast computations of frequency-response functions, based on full solutions at only a few master frequencies (even only one). A mathematical method, studied and developed in the EU project TAYLOR, now being further refined. The benefit is to reduce greatly the solution time for a complete narrow-band FRF.
- Tyre noise prediction, analysis and reduction: the EU project TINO, in which FEM/IFEM models of noise radiation, including tyre vibrations, shape, tread, *etc*, are studied.
- Domain decomposition techniques applied to structural dynamics and acoustics and automated parallelization tools (EU projects DOMINOS and SEP-Tools) – mathematical and numerical-processing methods to speed-up solutions and use new types of computer.
- Modelling flow-induced noise (computational aero-acoustics) using computational fluid dynamics and FEM/IFEM and BEM acoustic field models: the EU project ALESSIA, based on large eddy simulations in CFD, and researching other aspects of the methods.

8.2 Conclusion: future prospects

This paper has attempted to show the breadth of applications and methods in acoustic and vibro-acoustic modelling. The techniques available today are able to deliver tangible benefits in the product design process and in assessing acoustic performance. There are, of course, limitations. Overcoming these limitations is the challenge to today's researchers: the ever-increasing power of computers provides an environment in which advanced numerical methods can be exploited, but it is still no easy task to devise models, solution algorithms and user interfaces which make these methods available to regular industrial users and effective in real applications. Developments can follow the twin tracks:

- Increased speed and automation of the solution
- More complex models taking more physical effects into account

There are many issues on which partnerships of researchers, software developers and industrial users can collaborate.

S682/002/99

Visions and realities in vehicle noise and vibration

P VAN DER LINDEN, K WYCKAERT, and **C F McCULLOCH**
LMS International, Leuven, Belgium

1 SYNOPSIS

The marketplace and government legislation demand refined vehicles with reduced environmental impact. These goals have to be achieved in a highly-competitive market, where reduced cost, reduced weight and the use of advanced materials also play major roles. The vision of a refined vehicle has to be achieved through 'smart' design, implemented in practical development tools.

The paper presents methods for design analysis and refinement, using a hybrid or synthesis of modelling and test-based techniques, which can be used in a smart design process. After an overview of diagnosis methods for structure borne sound and airborne sound - experimental and simulation modelling methods, including Transfer Path Analysis, Airborne Source Quantification, FRF-Based Sub-structuring and Modal Sub-structuring - some case studies are presented to illustrate the procedures.

The approach can be both top-down (from vehicle behaviour to required module behaviour and parameters) and bottom-up (synthesising the vehicle NVH from module tests and numerical models). These methods enable smart design of vehicle modules, sensible module target setting and careful and smart module integration into the complete vehicles of the future.

2 INTRODUCTION: VEHICLE ENGINEERING PROCESSES

Vehicles are assemblies of modules: engine, powertrain, tyres, dashboard, steering system, body floor structure, *etc*. Nowadays, these modules are developed in parallel at multiple locations and often for more than one vehicle platform. The process of integrating modules in a complete vehicle development has also become a parallel rather than sequential development activity. Economic necessity demands the creation of new vehicles in shorter timescales, with key engineering decisions taken later in the process. Consequently, much of the engineering design work is done in parallel, on different modules, rather than as a sequential activity, and there is little time for changes to correct problems identified late in the process.

The noise and vibration (NVH) properties of the assembled vehicle will appear shortly before production starts, on the first realistic prototypes. Under the pressures of normal costs and product release targets, this does not leave much scope for further NVH development. The NVH quality therefore relies strongly on the specification and assessment process of the modules as well as on the early part of the integration process. A graphical view of the overall vehicle NVH process (as well as other design issues) is shown in Figure 1.

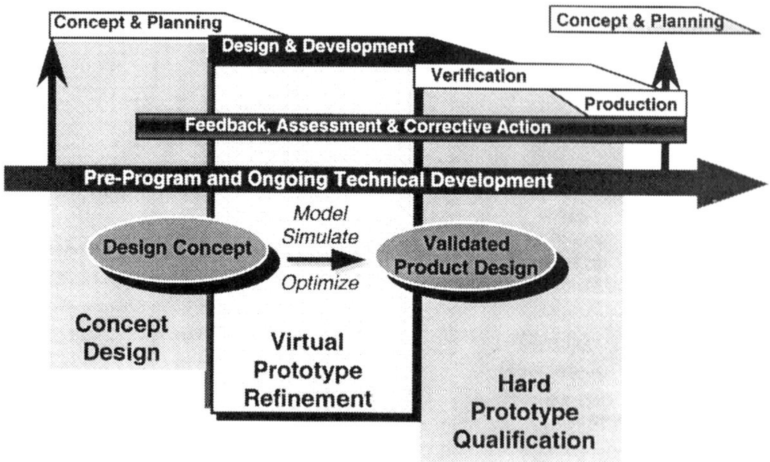

Figure 1: Vehicle programme and activities

The feasibility of NVH development on the modules and of early assessment of the NVH impact of proposed packaging and module interfacing solutions is an issue, both on the technical as well as on the organisational side. The decision-making processes in module target-setting, design and interfacing can be enhanced by recent improvements of sub-structuring and source-transfer system techniques.

The accuracy of assessment is improved by these techniques in two ways:

- Top-down, from vehicle behaviour to required module behaviour and parameters. The source transfer system techniques for airborne and structure borne effects can be used in many situations to derive adequate module NVH requirements from whole-vehicle NVH requirements.
- Bottom-up, synthesising the vehicle NVH from module test and numerical models. Modal Sub-structuring and Frequency-response-function Sub-structuring can be used in those situations where the module's dynamics are strongly coupled to its surrounding structure. (In such a case, individual module requirements from a top-down approach

will not necessarily ensure good assembled system behaviour). Sub-structuring can produce estimations of the NVH effect by combining data from existing and new modules or connections.

3 METHODS FOR NOISE AND VIBRATION DESIGN

3.1 Background to developments - EU programmes

As well as on-going development activities and consultancy-driven research within LMS, methods described here have been developed or refined in the framework of European Union research programmes, such as ESPRIT and BRITE-EURAM. By close cooperation with end-users the various techniques have been refined and validated in industrial situations. Practical implementation of the transfer path analysis and sub-structuring approaches has proved to be straightforward in some situations and difficult in others. Special instrumentation and routines were developed to support their application. The main programmes are:

- BRITE/DIANA: resulted in the structure-borne Transfer Path Analysis (TPA) methodology and software, for excitations from both drive-line and road inputs
- BRITE/PIANO (pass-by-noise analysis of heavy trucks): resulted in the airborne transfer path analysis methodology (airborne source quantification, ASQ) for radiation and complex reflection and isolation
- BRITE/VAMP (medium frequency modeling): established the basis for FRF-based sub-structuring
- EUREKA/EUROTRANS: resulted in the principal component analysis methodology, now a basic component of many transfer path analysis and other test-based methods
- BRITE/ASANCA: provided a validation of the structure-borne TPA methodology for aircraft
- BRITE/ANRAVA: refined structure-borne TPA for wheel suspension systems and road excitations
- BRITE/SOCRATES: refined FRF-based sub-structuring techniques and time-domain TPA methodology were developed.

3.2 Structure-borne and airborne transfer path analysis

3.2.1 Principles of structure borne or airborne transfer path analysis

A separation is made through the vehicle, creating pairs of interfacing subsystems. At the interface one subsystem is excited and produces relevant noise and/or vibration, and the total NVH response is expressed as a sum of contributions. The contributions are grouped according to their interface excitation: contact forces, contact acceleration, noise-radiating surfaces and surfaces subject to irradiation. Sound pressure and vibration acceleration responses result, or contact forces or particle velocity in a sound field can be found. The source transfer formulations are spectra and frequency response functions but can be transferred to the time domain in many cases. For the excited sub-system linear behaviour is assumed, which therefore only covers linear or weakly non-linear sub-system transfer. Because of the linearity, reciprocity also holds and this can be exploited in tests and simulation to gather the required transfer data.

3.2.2 Airborne radiation

Arbitrary vibrating components with arbitrary surface vibration fields and/or pulsation from nozzles or apertures can be discretised by omnidirectional volume acceleration point sources: $P_r = \sum_i H_{ri}^{P\dot{Q}} \dot{Q}_i \quad (Pa)$ where $\dot{Q}_i$ = volume acceleration of point source or patch i (m^3s^{-2}) and $H_{ri}^{P\dot{Q}}$ = transfer function of the sound pressure at r due to point source i ($Pa.m^{-3}s^2$) - see Figure 2.

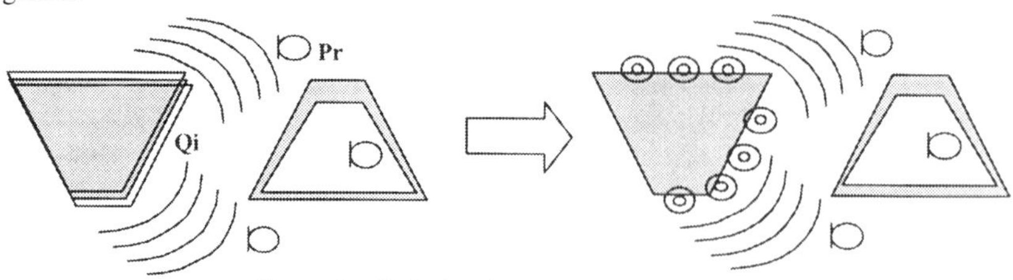

Figure 2: Substitution point source principle

This discretisation is valid in the sound field as long as the distance from point source to point source on the object is well below a half wavelength in the air. The equivalent point source volume acceleration of a surface patch can then be calculated by simple integration of surface normal accelerations over the patch. In many situations the dimensions of the surface patches or point source spacings can be chosen well above half an acoustic wavelength with an acceptable loss in accuracy and information density. The point source principle is used in models, and practical measurement methods exist for the entire audio frequency range.

3.2.3 Airborne irradiation

If a subsystem is subject to a pressure field, the sound pressure exciting a subsystem is discretised by concentrated forces and a sound pressure output is a sum of contributions: $P_r = \sum_i \overline{H}_{ri}^{PF} . F_i \quad (Pa)$ where F_i = surface normal force at patch i (N) and $\overline{H}_{ri}^{PF}$ = patch-averaged transfer function of sound pressure at r due to surface normal force i (m^{-2}). The substitution forces exciting a surface patch are calculated from the surface pressure using:

$F_i = \int_{S_i} P_i ds \quad (N)$ where P_i = pressure on surface patch i (Nm^{-2}) - see Figure 3.

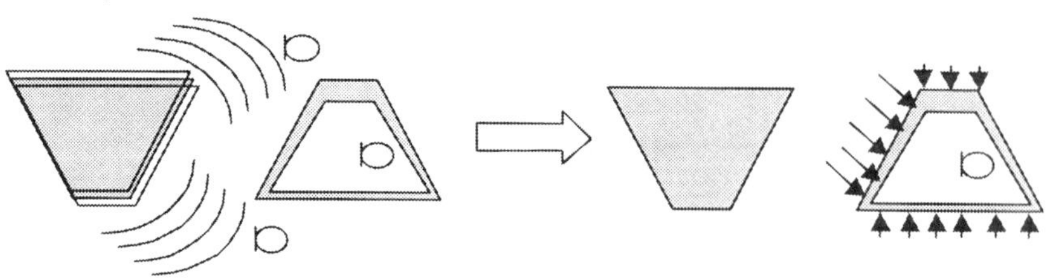

Figure 3: Substitution force principle

The irradiation can equally well be formulated in terms of pressure, and transfer from pressure to output. In either case, the patch dimensions should be well below a half wavelength in the air for a correct discretisation of the pressure excitation. For the correct vibro-acoustic transfer, the criterion is half a bending wavelength on the surface of the excited subsystem. However, larger patches can be used in some situations.

3.2.4 Structure-borne forces

Excitation by structural contact between objects can be seen as a source transfer system. Contact forces, moments and stresses are discretised by concentrated substitution forces and an output response acceleration is the sum of contributions from individual contact forces. This procedure is widely used and is covered in the literature. Several methods exist for identifying the substitution contact forces.

3.2.5 Structure-borne accelerations

An alternative to source transfer uses the velocity or acceleration at the structural contact as discrete excitations, again with an output acceleration as the sum of contributions. This procedure is not often applied, but an example is given later in this paper.

3.3 Sub-system assembly methods

3.3.1 General principles

The assembly of sub-systems is performed stepwise, adding sub-systems one by one to build up a chain from initial excitation to output response.

The NVH of assemblies can be analyzed by detailed numerical models of all acoustics and structures. This may not always be practical because of model size, and accurate models of all sub-systems in an assembly chain are not always available. Reduced representations of the sub-systems can be used: modal models, FRF models and statistical energy flow models. These reduced, linear, models can be sufficient for the understanding of the full system behaviour and for the evaluation of modifications.

It is also possible to combine test databases and simulation models of individual sub-systems to study the behaviour of the assembly. Test databases will, in general, only provide reduced models of sub-systems, but these modal, FRF or statistical test models can often be used to good effect in an assembly.

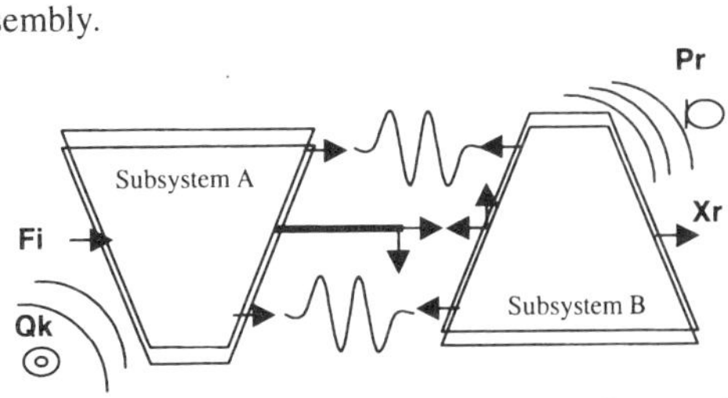

Figure 4: Sub-system assembly: structural and acoustic excitation, structural and acoustic responses

3.3.2 Modal synthesis

Modal sub-structure assembly (synthesis) has been implemented, applied and documented for a long time. Both constrained and unconstrained interface formulations have been described for both FE models and physical tests. For test-based sub-system models, modal synthesis normally starts from modal data for unconstrained (free-free) sub-systems. To create true rigid boundary conditions for test objects is often not feasible. Both springs and rigid connections can be defined between the sub-systems, after which the assembled system modes are calculated. This synthesized modal model can then be used for forced response calculations to determine critical output responses. Even acoustic cavities can be described by experimental and (or numerical) modal analysis, but modal synthesis of cavities and structures is not well developed.

If a modal database of sub-systems with unconstrained coupling interfaces is used for synthesis, then truncation errors are encountered. The stiffness at the interfaces is poorly represented in these modal databases. The truncation effect can be reduced by inclusion of modes at frequencies well above the frequency range of interest. The other way to address this issue is to use constrained mode models. Finite element based modal analysis can provide accurate modal databases of constrained systems.

An advantage of modal models is that they usually provide a very compact description of sub-systems, whilst still allowing a high number of interface, excitation and response degrees of freedom.

3.3.3 Frequency-response-function synthesis

Individual sub-systems are described by their frequency response functions (calculated or measured) and the assembled system FRFs are derived. The response of a sub-system B when connected to an excited sub-system A is calculated in two steps using the sub-system matrices: first the interface forces are calculated and then these are used to calculate the output response. The required matrices consist of frequency response functions of the isolated sub-systems in free-free conditions. Rigid connections can be modeled, and springs connecting interfaces. Rubber mounts (springs) are described as a separate matrix in the formulation, which allows an efficient analysis of variations in coupling stiffnesses. A similar formulation is possible if both excitation and output responses are on sub-system B, or both on A, and so on.

Because of the inversion of the coupling matrix, highly-accurate FRF data are required. When using test databases, a limited number of effective coupling DOFs can be treated. When using FE models to generate FRFs through modal synthesis (the most common procedure) modal truncation errors can be critical. This can however be corrected by adding residual modes, calculated from the difference between the modal forced responses and direct frequency responses.

3.3.4 Statistical energy analysis

Noise and vibration synthesis at higher frequencies cannot use modal and FRF synthesis techniques, because the modal density and modal overlap in the sub-systems are too high, but Statistical Energy Analysis models are well-suited to this frequency range. SEA works with averaged energy responses in the different sub-systems and the power flow through their assembly: for each sub-system, energy can be lost through internal dissipation (internal loss factor) and through coupling interaction with other sub-systems in the assembly (coupling loss factors). Contrary to the situation for modal and FRF synthesis, the coupling loss factors are not independent of sub-system properties, but will depend on both sub-systems at their interface. Structural and/or acoustic energy is input to the relevant sub-systems, as power, or as forces, accelerations, pressures and volume accelerations. Non-power inputs require that additional data are known, for sub-system impedances, but in principle they provide the possibility to combine transfer path analysis with SEA.

Synthesis based on energy methods will be effective in the frequency range where a high modal density exists in all relevant subsystems. Automotive applications necessarily focus on body development in medium and high frequencies: body structure, damping layers, interior trim, engine and interior acoustics.

3.4 Sub-system definition and requirements

3.4.1 Principles and practicalities

NVH responsibility is distributed because component and major system development is distributed. Suppliers provide sub-systems for vehicle integration - engines, rubber mounts, interior trim, tyres, exhaust systems, dash panel... Different departments at vehicle manufacturers are responsible for individual modules: body-in-white structure, wheels/suspensions, steering systems, heating/ventilation systems... Almost every sub-system on a vehicle has NVH-related excitations, external/internal, airborne/structure borne. Almost every subsystem will also contribute something to whole-vehicle NVH.

The organisational responsibility may likely define the interfaces clearly. If this is not the case, then the degree of coupling between the subsystems should be considered when defining the separating interfaces, because weakly-coupled sub-systems have distinct advantages for NVH development.

3.4.2 Degree of coupling between sub-systems

A sub-system is considered predominantly decoupled if its own transfer behaviour changes little when there are changes in surrounding sub-systems. At structural mounting interfaces this requires that the point accelerance (or structural input impedance) of the sub-system at the relevant interface DOFs differs greatly from the input accelerance of the surrounding sub-systems. The DOF by DOF evaluation of the accelerance ratio will usually determine between strongly- and weakly-coupled interfaces.

For a weakly coupled situation, sub-system A can be far stiffer and/or heavier than sub-sytem B at the interface: sub-system A will then respond almost as if in free-free conditions for this DOF and sub-sytem B will respond almost as if clamped. Alternatively, sub-system A can be far more compliant and/or lighter than sub-system B at the interface: sub-system A will then respond almost as if clamped and sub-system B will respond almost as if in free-free condition. With airborne noise the degree of coupling can also be evaluated, both for exposed surfaces of sub-systems, subject to irradiation, and for exposed surfaces radiating noise. A large difference in acoustic impedance is required for weak coupling. Acoustic impedances can be calculated from the geometry of a small cavity, an open region can be assumed to have an impedance of the same order as free wave propagation, and for structural radiating/irradiated surfaces from the patch geometry and the acoustic and structural bending wavelengths.

For automotive applications, the acoustic impedance of the air is in most cases far smaller than the acoustic input impedance of structures. In other words, structures will vibrate freely, unaffected by air loading. Surface vibration is also hardly affected by large changes in the surrounding sound field. Most apertures or nozzles, producing noise (exhaust nozzles, intakes, ventilation openings...) often also have a higher impedance than their surrounding sound field. Only acoustic absorption materials cannot be considered weakly coupled to the air: the sound field is changed by introducing absorbent. For engineering purposes, an accelerance or acoustic impedance ratio larger than 10 dB can be used as a criterion. This condition should be satisfied before and after modifications, and for all relevant interface DOFs within the frequency range of interest.

3.4.3 Handling strongly-coupled sub-systems

Sub-systems with input accellerances and/or acoustic impedances at the relevant interface DOFs and/or surfaces which are of the same order of magnitude, are strongly-coupled. They will also be subject to excitation and produce output responses as weakly-coupled sub-systems do, so can also be described with all transfer path analysis techniques. However, for strongly-coupled sub-systems a modification on one sub-system, will not only modify its own transfer behaviour, but will also significantly modify the response of surrounding sub-systems. Therefore, a reduction of the transfer from input to output of a strongly-coupled sub-system will not always result in a reduction of that sub-system's output or of the assembly output.

For strongly-coupled sub-systems, assessment of design modifications or variants can only be judged from assembly analysis, using one of the synthesis techniques, whereas for weakly-coupled subsystems, local sub-system input-output behaviour should be sufficient for the evaluation of possible modifications.

3.4.4 Rubber-mounted sub-systems

Ideally, a system is separated at compliant rubber mounts, because this usually gives weak coupling of the sub-systems. An easy analysis of the effects of mount stiffness modifications is also possible. For compliantly-mounted sub-systems, two types of separation are possible:

- between the active sub-system and the mounts: the acceleration is the more stable excitation quantity;
- between the mounts and the supporting/passive sub-system: the force is the more stable quantity.

3.4.5 Test-beds for isolated sub-systems

A test-bed for a sub-system in isolation should take into account the interface situation of the sub-system on the vehicle. If a component is surrounded by others with higher accelerances, or by compliant mounts, then free-free test conditions are the most appropriate pure boundary condition to use on a testbed. For example, a radiator/fan assembly without its rubber supports. If a component is more flexible than its surroundings on the vehicle, then the best approximation to rigid boundary conditions is desirable. For example, tyres. If a component is strongly-coupled to its adjoining sub-systems on the vehicle, then the synthesis technique to be used will determine the required boundary conditions for tests on the component in isolation. If one expects the surrounding sub-systems to remain unchanged, than the vehicle situation can be mimicked (which is not uncommon in practice). Calculations on hybrid or full simulation models can follow the same rules as for the testbeds for the best pure boundary conditions and loads to chose in the modelling.

4 TARGETS IN SUB-SYSTEM DEVELOPMENT

4.1 General principles in target setting

A major use of transfer path analysis and synthesis techniques is the definition of sub-system targets on the basis of the complete vehicle NVH targets (although sub-system targets may be defined by non-technical factors such as product image).

If a sub-system is predominantly decoupled, both structurally and airborne, then it becomes possible to project the vehicle NVH targets downwards on to the sub-systems using transfer path analysis as a basis. This is a fairly stable process with respect to changing components and interfaces. Ideally, each sub-system should be given a target range for the transfer between its interface excitations and its outputs, or, for sub-systems which are internally excited, a target range for the allowed output levels at the interfaces.

It must be noted that if sub-systems are strongly coupled, then isolated sub-system targets can quickly become incorrect as surrounding components and interfaces are changed during development.

4.2 Cascading targets for sub-systems

NVH targets are usually defined for several vehicle conditions which highlight different phenomena, like wind noise, road-induced vibration and noise, engine noise, idle shake, *etc*. This tends to reduce the number of sub-systems and transfer paths involved in each NVH effect.

Potentially, many paths contribute to the perceived NVH. One could assume that in a balanced situation all interface DOFs and/or surfaces (paths) will equally contribute to the NVH quantity of the vehicle, which will tend to give very severe requirements on each path. Alternatively, one could assume that in critical situations only one path is relevant, but to allow each path to reach the total target will probably lead to a problem. A compromise has to be found through the choice of a safety factor: for example, diving the total NVH target by the square-root of the number of paths can provide a reasonable target for each path.

The relative phase between paths can also affect the total NVH. In some cases, for low frequency phenomena (*eg* idle boom/shake) it could be necessary to assume in-phase contributions. In most practical situtations, the phase relations between path contributions are unknown and/or highly condition-dependent - in which case a random phase assumption is probably best.

In the next step, for each path an excitation target and a transfer target are required. These targets can only be set on the basis of references. Newly-developed sub-systems will usually not vary much from their predecessors or reference sub-systems. For example, car and truck bodies currently create on the average 0.015 Pa of interior noise for 1 N excitaton force at attachment points (depending on the nature of the attachment point) but similar-sized vehicles vary ± 5 dB in vibro-acoustic transfer for similar attachments.

4.3 Guidance for the development of coupled sub-systems

Stable targets can be set for a chain of coupled sub-systems. The chain starts at a weakly-coupled interface and ends at a weakly-coupled interface. Examples of strongly-coupled sub-system chains include:

- Vehicle interior noise from low-order engine vibrations – which can be translated into target ranges for allowed acceleration levels on the engine and transmission brackets for rubber mounts. To translate this into a required structural transfer of the engine and required input accelerances of the interface between the engine and the transmission will require a coupled sub-system analysis. The chain will therefore have to cover engine, auxilliaries and transmission, up to the rubber mounts.
- Tyre non-uniformity/unbalance – for which the tactile steering wheel vibration is relevant. The chain starts at the tyre excitation on the rim, goes through wheel, suspension system, steering system, body structure, dashboard carrier, up to steering wheel. Unfortunately, this covers almost the entire vehicle.

The effects of a new or modified heavily-coupled sub-system can be judged by modeling the new sub-system surrounded by others, assuming they are unaltered (one exploits the fact that often only relatively small modifications are possible, due to technical and financial reasons). The surrounding sub-system models can come from experimental modal analysis, measured FRF matrices or validated FEM/BEM modules. The output at the interfaces of the chain can then be judged relative to the path targets for the chain.

5 APPLICATIONS

5.1 Vehicle pass-by noise

Pass-by noise is the most critical aspect in truck NVH development. The pass-by noise in the frequency range from 300-3000 Hz is expressed as a sum of contributions, which are the products of airborne source strengths and acoustic transfers.

The major driveline components, intake system and exhaust system are each characterised by the volume acceleration. The volume acceleration can be identified for each smaller component (exhaust shell, tail pipe, oilsump,...) or it can be expressed for the full system (complete engine, exhaust system, intake system). The vehicle encapsulation, shielding, absorbtion *etc* is taken into account in the acoustic transfer functions. The vehicle acoustic transfer can be averaged over the complete engine, or also expressed for each component (oilsump, distribution cover, *etc*).

In this case study, only two levels of subsystems or modules are considered: the active engine, exhaust *etc*, and the passive encapsulation, cabin, *etc*. These subsystems are well decoupled at their interface and it is possible to translate the allowed maximum dB(A) down to allowed volume acceleration spectra and allowed acoustic transfer curves (Figures5).

The volume accelerations of each active subsystem can be measured on a fully-operational vehicle, or these can be determined on an engine testbed. The use of data from testbed measurements enables the evaluation of different gear ratios on the pass-by noise. The acoustic vehicle transfer function is measured efficiently by means of reciprocity. The acoustic transfer is fully-determined by shielding encapsulation and cabin/chassis diffracting effects. The measurement does not require a functional driveline. Several variant vehicle configurations can be assessed for their impact on pass-by noise, without the need for repeated realistic pass-by tests in combination with functional driveline variants.

Another advantage of this modular approach to pass-by lies in the use of numerical modeling for the design guidance. Full pass-by noise modelling is currently not possible, but numerical modelling of a critical engine part or of the sound field around one of the critical parts can be feasible.

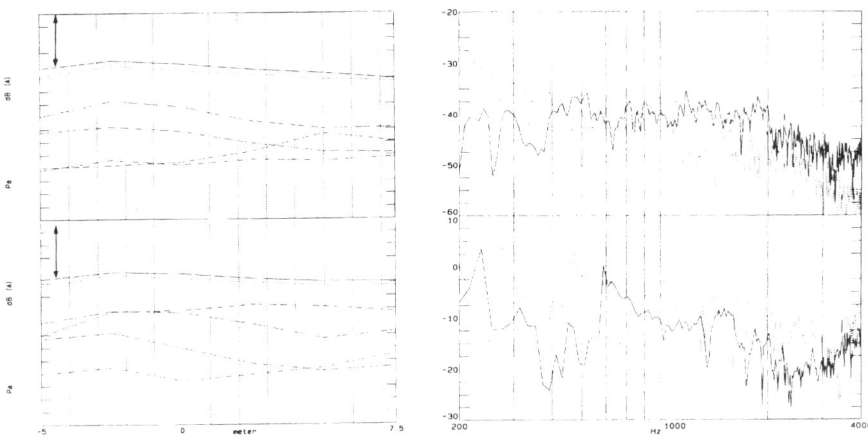

Figure 5: *Left*: Overall dB(A) *vs.* position; total measured (solid), total identified (dots) and some contributions examples of parts of the engine, exhaust and gearbox (dashes and dots), upper window right side, lower window left side; *Right*: Narrow-band acoustic transfer (upper window) and narrow-band volume velocity autopower spectrum (lower window) of part of the exhaust (solid) and part of the engine (dashed)

(by courtesy Renault Vehicules Industr.)

5.2 Rear axle interior road noise

Structure borne transfer of road-induced vibration into the interior is important for NVH development, but is also difficult because of the complexity and variability of the road excitation and because of the relatively complex dynamics of the system involved.

In this example, an FE model of a twist beam rear axle is combined with measured FRF data of a fully trimmed body to evaluate the road noise. There are two subsystems, the trimmed body and the rear wheel suspension, with a relatively strong dynamic interaction.

One could argue that in this particular example of the suspension body interface, shown below, a fairly weak coupling exists, but in general this is currently not the case for wheel-suspension-body interfaces. Even though body structures are becoming stiffer and wheel suspensions more compliant, in many cases a decoupled situation cannot be reached whilst retaining good vehicle stability.

The body is one sub-system, the rear-axle including the hubs, brakes and rims are the next sub-system and the tyres represent a third sub-system. The tyres are considered constant force sources, which is justified from the input accelerance difference between the wheels and the tyres (from 60 Hz upwards). The rear axle is driven by forces, identified from measurements. In this case, the wheels belong to the rear axle. This would allow a wide range of modification on the wheel suspension system, including the wheel assemblies.

Other approaches to define the interfaces are also possible. One could include the tyres at the rear axle and consider acceleration excitation at the road interface. The excitation could then be considered stable down to 0.5 Hz. Alternatively, the wheel suspension could also be driven by measured accelerations at wheel centre. This means that the tyres, wheels, hubs and brakes are the third sub-system, which would then be considered a constant velocity source from 200 Hz upwards for this type of rear axle.

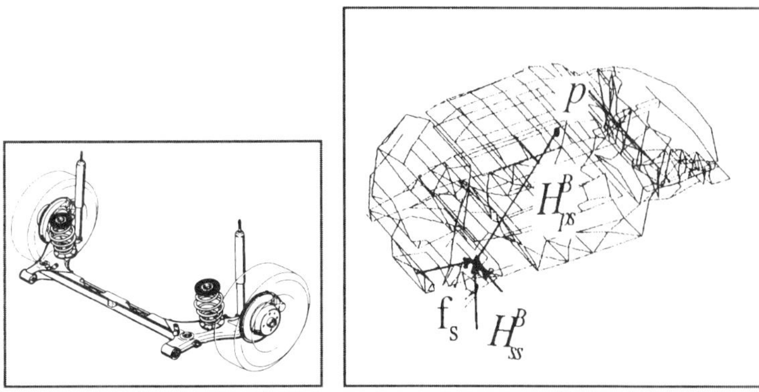

Figure 6: Schematic representation of the coupling procedure
(by courtesy RNUR)

The behaviour of the rear axle is assessed using FRF sub-structuring. The body interface forces are calculated from FRF matrices, and the formulation explicitly allows insertion of the mount stiffness as a separate subsystem. The interior noise is calculated from the coupled body interface forces in combination with the vibro-acoustical FRF.

The FE model of a new, modified, rear axle design is used to calculate new transfer matrices, which are coupled to the matrices of the existing body to evaluate the proposed design. The comparison is made at the level of the interior noise, as can be seen in the Figure 7 (right).

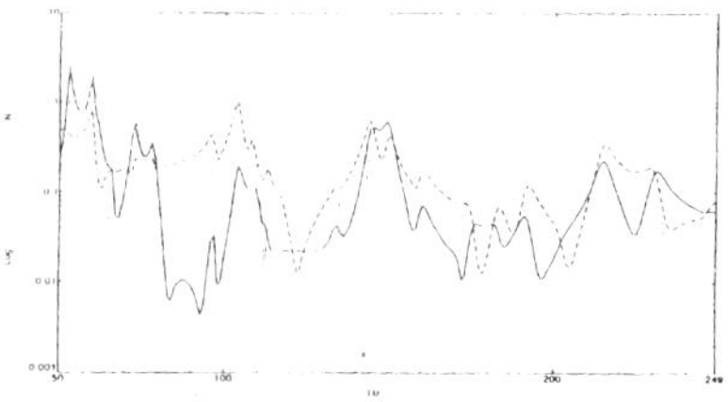

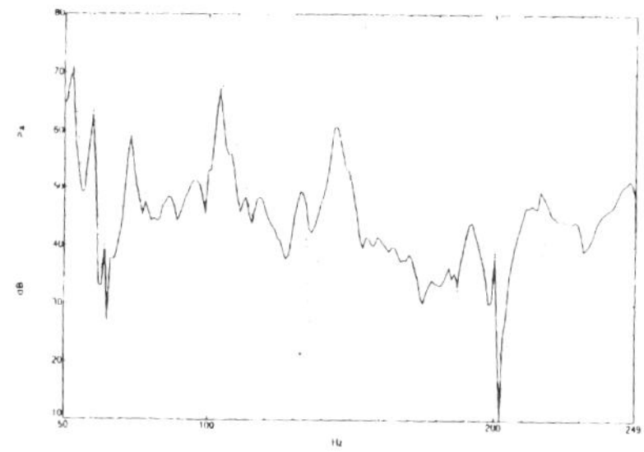

Figure 7:

Coupled force transmissibilities **Calculated interior noise before and after modification**

(by courtesy RNUR)

5.3 Oilsump noise radiation

On modern engines the oilsump and other covers have become the main noise-radiating components both for interior noise and for pass-by noise. Engine development for airborne noise is usually judged on the basis of averaged sound pressure at 1 metre distance in anechoic conditions. The oilsump noise emission in this case is therefore analysed by sound pressure, but this can be easily translated into volume acceleration following the transfer path procedure (as in the truck pass-by noise example).

The oilsump is defined as one subsystem, the rest of the running engine is defined as the second subsystem. Because the engine structure tends to be both far stiffer and heavier than the oilsump, we have a decoupled situation for many engine-oilsump combinations. The engine will generate a constant acceleration excitation, and the squared averaged sound pressure, at 1 metre distance from the oilsump alone, is calculated from the squared averaged accelerations at the engine interface. The engine acceleration level is easily determined from accelerometer measurements on a testbed. The oilsump acceleration acoustic transfer can be measured on an oilsump alone (see Figure 8, left).

The transfer can also be calculated on the basis of a structural finite element models combined with an acoustic boundary element model for the sound pressure calculation.

Oilsump designs are often highly variable, because of adaptations to vehicle packaging, efforts to reduce the noise, costs, *etc*. The modular approach can allow a more efficient validation of the designs, or a target for the oilsump acceleration acoustic transfer function can be derived from engine noise emmission targets.

Oilsump transfer measurements set-up

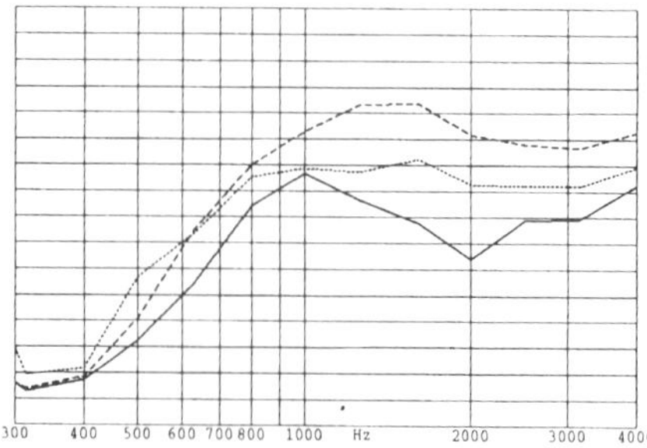

Figure 8:

Acceleration acoustic transfer of oilsumps of different materials

(by courtesy DaimlerChrysler)

6 CONCLUSIONS

- Transfer path analysis on both airborne and structure borne phenomena can make it feasible to perform independent NVH development on sub-systems, namely the modules consisting of powertrain, mounts, sub-frames, axles, body, tyres, *etc*.
- The impedance or accelerance of the interfaces determines whether a sub-system can be developed using transfer path techniques and sub-system targets, or if synthesis techniques are necessary.
- A careful selection of the sub-system separation can ensure weak structural coupling at the interfaces, which simplifies the sub-system development. For airborne phenomena most sub-systems can be considered weakly coupled.
- Synthesis techniques can be used in the development of structural sub-systems that are strongly-coupled, but these methods require relatively large and accurate sub-system models (FE, FRF, Modal, SEA, ...).
- Databases of excitations (forces, volume accelerations, *etc*) - both external and internal to sub-systems - are essential for modular NVH development.

S682/003/99

Understanding and controlling acoustic effects of outdoor surfaces

K ATTENBOROUGH
School of Engineering, University of Hull, UK

1 INTRODUCTION

Destructive interference between sound travelling directly to source and receiver and that reflected, with frequency-dependent phase change, from a finite impedance ground surface produces attenuation in excess of that resulting from wave-front spreading and atmospheric absorption. This is often known as ground effect. For source and receiver heights of the order of 1m over grassland, ground effect is manifested by extra attenuation in the frequency range 200 Hz to 800 Hz. Under calm, acoustically-neutral conditions the extra attenuation over 1 km due to ground effect can be as much as 35 - 40 dB in the 250 Hz octave band [1]. Measurements obtained with a fixed jet engine source reveal the large effect at low frequencies (63 Hz and 125 Hz octave bands) that results from the presence of thick snow [1]. A similar low-frequency ground effect is associated with propagation through mature forest stands [2,3]. Atmospheric turbulence reduces the coherence between direct and ground-reflected sound and hence reduces the magnitude of the ground effect. Atmospheric refraction has an important influence also. Ground effect is increased under inversion or downwind conditions and reduced under upwind and lapse conditions. Predictions of outdoor sound propagation often take ground effect into account through estimates or measurements of the normalised surface impedance of the ground. So far the emphasis has been on characterising the given ground surface in this way rather than on deliberately engineering the surface for optimum noise reduction. This is in contrast with the effort that has been devoted to designing pervious road surfaces and absorbent coatings for noise barriers. The prediction of the effects of porous road surfaces is complicated by their influence on the generation of tyre/road noise as well as on subsequent propagation. It has been suggested that it is possible find ground parameters, including flow resistivity, porosity and roughness, that optimise ground effect and that the optimised ground effect is not affected greatly by weak refraction [4]. Such optimisation is predicted to be most effective for low frequency sources near to the ground. As is well known, flow resistivity is the most important parameter. For a given source spectrum and range, ground effect is predicted to increase rapidly for decreasing flow resistivity and is relatively insensitive to flow resistivity once the optimum has been reached. Apart from designing porous road surfaces, other opportunities for noise reduction outdoors include exploitation of railway ballast and cultivation of land around major noise surfaces.

2 FACTORS THAT INFLUENCE GROUND EFFECT

Ground effect is sensitive to the acoustical properties of the surface which depend on porosity and flow resistivity (or air permeability). Soils have volume porosities of between 10% and 40%. Snow (porosity of around 60%) and porous road surfaces (porosities between 30 and 40%), have fairly low flow resistivities. A wet compacted silt soil surface will have a rather high flow resistivity. The thickness of the surface porous layer is important and whether or not it has acoustically-hard substrate. The roughness of the surface is important. Small-scale roughness of the ground, such that the mean roughness height is less than the sound wavelengths of interest, influences its apparent surface impedance [5,6]. If the surface is otherwise acoustically-hard, the presence of roughness serves to 'soften' it and alter the sound attenuation at near-grazing incidence. As the mean roughness dimension increases the relative contribution of incoherent scattering increases and the ground effect is reduced. Deliberate corrugation of the surfaces at the sides and in the central reservation of a motorway is predicted to give extra attenuation comparable with that obtained from pervious road surfaces [7].

Potentially, farming practices such as sub-soiling, disking and ploughing have important influences on ground effect since they change both the roughness and surface properties (including the air permeability) of the ground. Aylor [8] noted a significant change in the excess attenuation at a range of 50 m over a soil after disking without any noticeable change in the meteorological conditions. Figures 1 and 2 show spectra of the difference in sound pressure levels received by vertically-separated microphones over ploughed and sub-soiled heavy (boulder) clay [9]. The microphones were at 1 m and 0.1 m heights and located 30 m from the Electro-Voice loudspeaker source (centre at 1.65 m height) on a sunny day with light wind. Such vertical level differences are surrogates for excess attenuation since they are independent of the source spectrum.

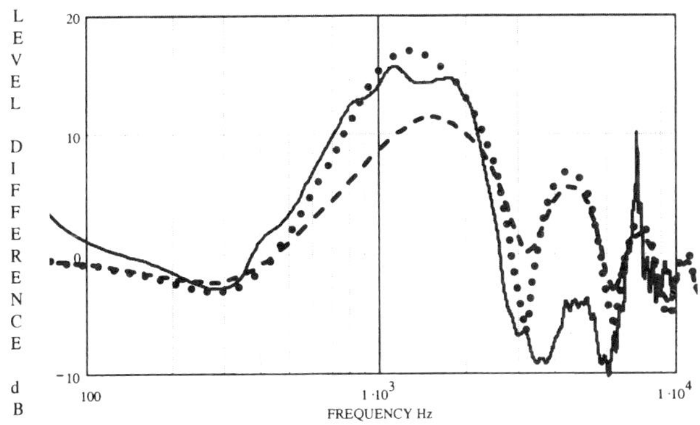

Figure 1

Average level difference spectrum obtained at a range of 30 m over sub-soiled ground (solid line). Predictions assume either the smooth surface of a slit pore medium with flow resistivity 80 kPa s m^{-2}, porosity 0.4, tortuosity 2.5 and thickness 0.08 m (broken line) or a rough surface with these parameters plus 0.05 m high close-packed but partially-random (randomness factor 0.75) semi-cylindrical roughness (dotted line). Turbulence is characterised by $<\mu^2> = 10^{-5.75}$, $L_0 = 1$ m.

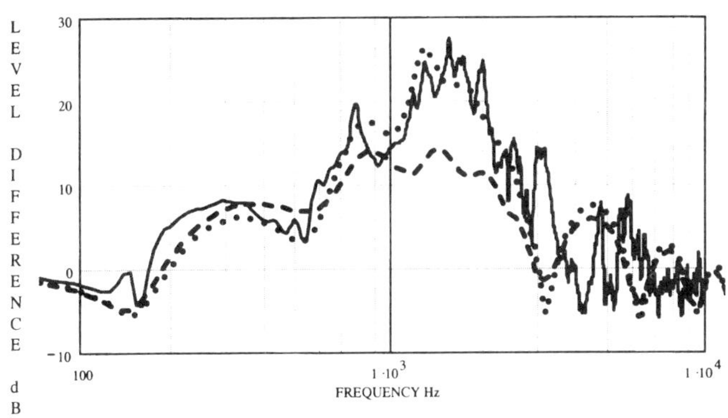

Figure 2

Average level difference spectrum obtained at a range of 30 m over ploughed ground (solid line). Predictions assume either the smooth surface of a slit pore medium with flow resistivity 10 kPa s m^{-2}, porosity 0.4, tortuosity 2.5 and thickness 0.175 m (broken line) or a rough surface with these parameters plus 0.08 m high, close-packed, semi-elliptical (eccentricity 1.6) roughness, (dotted line). Turbulence as for fig. 1.

There are clear differences in the ground effects. Figures 1 and 2 also show theoretical predictions for the level difference spectra at 30 m. Although some of the change in ground effect between the two conditions will result from differences in flow resistivity, improved agreement with data is obtained by attributing different surface roughness to each ground.

This is consistent with the fact that sub-soiling produces less surface roughness than ploughing.

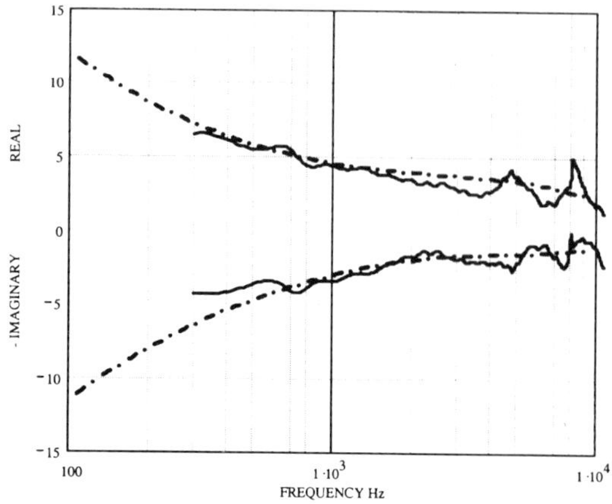

Figure 3

Normalised impedance data (solid lines) obtained from complex excess attenuation measurements over a sports field at the Open University, UK. Theoretical predictions (dash-dot lines) are for the impedance of a slit pore medium with flow resistivity 120 kPa s m^{-2}, porosity 0.4, tortuosity = 1/porosity and 0.005 m high semi-cylindrical roughness with mean spacing 0.04 m.

Recently, a method has been derived for deducing ground impedance from measurements of the complex excess attenuation spectrum due to a point source at short range [10]. Figure 3 shows the impedance of a sports field obtained in this way together with best-fit predictions assuming a rough porous surface. The sports field is flat and carefully maintained with mown short grass. The resulting impedance may be contrasted with that shown in Figure 4 obtained over established grassland at Silsoe in Beds (UK).

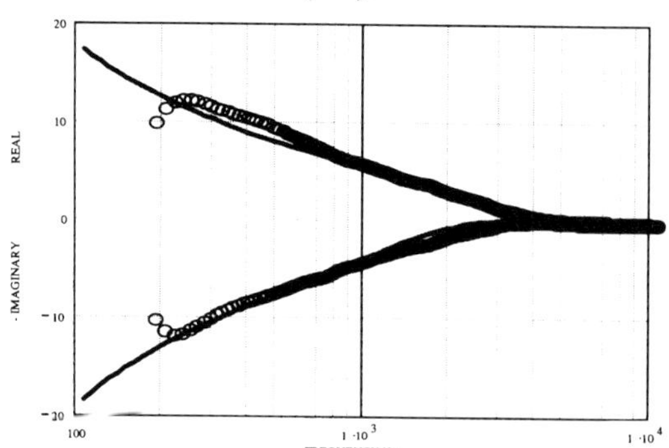

Figure 4

Normalised impedance data (open circles) obtained from complex excess attenuation measurements over established grassland at Silsoe, Beds, UK. The theoretical predictions (solid lines) are for the impedance of a slit pore medium, with flow resistivity 300 kPa s m^{-2}, porosity 0.4, tortuosity = 1/porosity. Partially-random (randomness parameter 0.5 [6]) close packed semi-cylindrical roughness with 0.02 m radius is assumed.

The data are the result of averaging impedance deductions from different complex excess attenuation measurements at adjacent locations. Note that, in this case, the resistance is somewhat greater than the reactance. The widely-used semi-empirical one-parameter model is not able to predict this behaviour. In general it is characteristic of a ground in which the flow resistivity decreases with depth. The data in Figure 4 is such that the reactance and resistance tend to zero at high frequencies. This is the predicted result of roughness. The best-fit roughness in this case is greater than that assumed for the sports field.

3 INFLUENCE OF MODIFYING GROUND ON A-WEIGHTED LEVELS

By using parameters corresponding to measured propagation data, together with theories for propagation from a point source over impedance ground, it is possible to explore the implications of ground treatment for the variation in A-weighted levels with distance. Figure 5 shows predictions of the ground effect relative to 13.5 m from the source for A-weighted traffic noise over a sub-soiled ground and a sports field. The corresponding impedances are given by the parameters identified in the captions for Figure 1 and Figure 3 respectively. Figure 5 shows several predictions of the variation, with distance, of ground effect relative to that at 13.5 m from a traffic noise source. The solid and broken lines are predictions of relative ground effect over simulated sports field and sub-soiled ground. They have been obtained by means of a semi-empirical formula for the effect of discontinuous impedance [13]. The assumed noise spectrum (shown in Figure 6) corresponds to free flowing traffic [14]. The assumed source height is at 0.5 m and the receiver is assumed at a height of 1.2 m with the source-receiver axis normal to the edge of the road. The dash-dot line results from use of the standard UK scheme (CRTN) for traffic noise prediction [12]. The dotted line corresponds to predictions of a formula, obtained by Hothersall *et al* [13] representing the best fit to the results of calculations with a boundary element code [15]. All of the predictions in Figure 6 assume 100% soft ground cover beyond 3.5 m from the traffic source, which is assumed to be above an acoustically-hard surface. The CRTN formula is based on a wide variety of data and the results of computer simulations. The predictions of relative ground effect over cultivated surfaces are practically identical and suggest significantly greater values than predicted by the CRTN scheme.

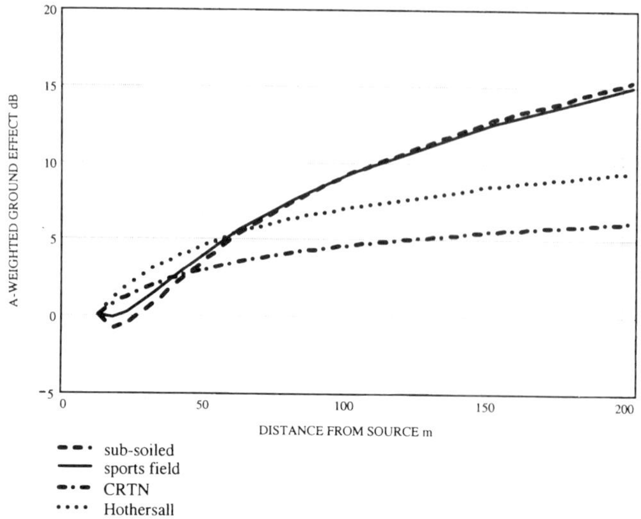

Figure 5
Predicted variation of ground effect relative to that at 13.5 m from the source (i.e. 10 m from the nearest edge of the road) for A-weighted traffic noise.

The boundary element predictions [13] assume a receiver height of 1.5 m and a ground impedance given by the semi-empirical one parameter model with effective flow resistivity of 250 kPa s m^{-2}. This corresponds to a higher impedance than those measured for the sports field or deduced for the cultivated soils. Hothersall et al [13] remark that their predictions are higher than the CRTN predictions since they do not include effects of turbulence or downwind conditions.

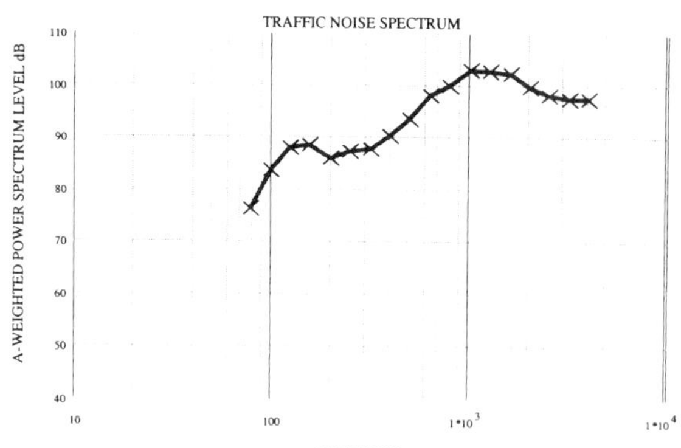

Figure 6
A-weighted traffic noise power spectrum deduced from SPL measurements at 20 m from edge of freeway [14].

Predicted effects of moderate turbulence, described by the mean squared refractive index and outer turbulence length scale, on A-weighted ground effect for traffic noise are shown in Figure 7. This suggests that such turbulence may reduce the predicted ground effect by up to 6 dB at 200 m. Nevertheless, there still may be a worthwhile enhancement of ground effect from use of an impedance that is lower than those assumed typical of soft ground.

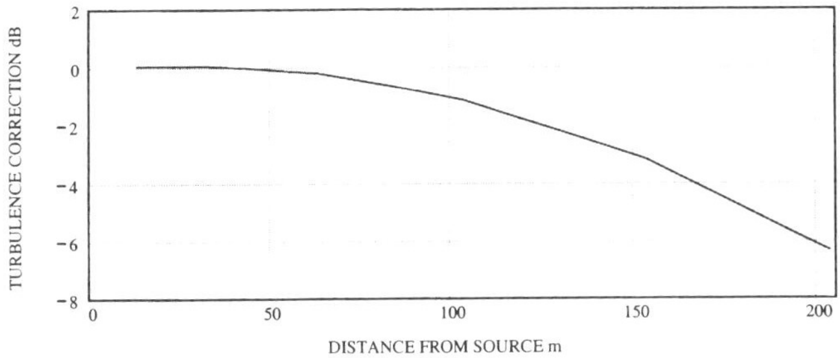

Figure 7

Predicted effect of moderate turbulence ($<\mu^2> = 10^{-5}$, $L_0 = 1$ m) on A-weighted ground effect for traffic noise.

4 OPTIMISATION OF RAILWAY BALLAST

Calculations have been performed using the boundary element method. With this method, the sound field is found by calculating the field at the boundaries of the volume that is considered, under the assumption of a homogeneous atmosphere. With the solution known at the boundary, the sound field can be calculated at an arbitrary point using analytical formulas for sound propagation in homogeneous air. The shape of the boundaries is, in principle, arbitrary. The code used in this study was originally developed to calculate the insertion loss of road barriers [15]. The code is two-dimensional. Only situations and sound sources can be modelled that show no variation along one axis. An infinite flat or profiled ground surface can be modelled. The acoustical impedance can be chosen independently for each surface element of the profile. Furthermore, it is possible to determine the surface impedance for each element using a dominant angle of incidence. The dominant angle of incidence is calculated from the relative position of the sound source and the surface element. The code has been extended to evaluate sound propagation over ballast. A slit-pore model has been used to calculate the acoustical properties of the ballast. The type of ballast found in the Dutch railway tracks [16] has been considered A draft European standard for unbound aggregates to be used as railway ballast [17] gives requirements for applicable materials that are similar to materials currently in use in the Netherlands. A number of stone size distributions is being used in ballast. A most common distribution involves stones diameter of 30 up to 63 mm. Smaller stone sizes can give rise to problems with the drainage, while larger stones can lead to problems with the construction and maintenance of the railway track. The porosity of a volume filled with stones depends on the distribution of stone sizes, on the rate of compaction, and on the typical shape of the stones. For railway ballast, the porosity is reported [16] to vary from 0.3 to 0.4. In this study, the flow resistivity is computed from [18]:

$\sigma = \mu / K$ Ns/m^4;

$$K = 0.00946(1 - \Omega)^{2/3}\left[\frac{0.904}{(1 - \Omega)^{2/3}} - 1\right]^2 D^2 \text{ m}^2,$$

where m denotes the dynamic viscosity for air and D the mean particle diameter in meters. The tortuosity is obtained from the relation [19]:

$$q^2 = \Omega^{-s},$$

where s denotes the shape factor. Values of tortuosity up to 2 are expected for ballast. Typically, there is a total ballast depth of approximately 0.5 m

The A-weighted emission spectrum for freight trains travelling at 100 km/h is used in the calculation of sound level differences. This spectrum is determined according to the Dutch Guidelines for the calculation of noise generated by railway traffic [20]. The normal ballast porosity is assumed to be 0.38. With a mean stone-size diameter of slightly less than 5 cm it follows that the flow resistivity is approximately 20 Ns/m^4. The tortuosity is assumed to be 1.622 and the ballast thickness is 0.5 m.

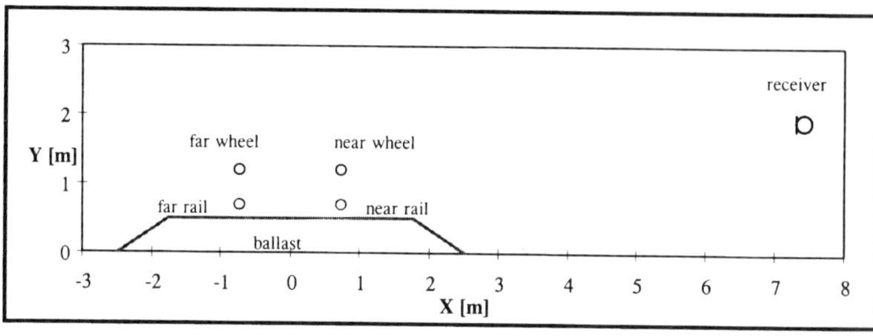

Figure 8

Geometry used in the calculations of acoustic effects of railway ballast

Fig. 8 shows the geometry assumed for the railway track. The nearest and furthest rails and corresponding wheels are considered separately. The rail noise sources are modelled as infinite line situated 0.2 m above the ballast surface. Noise radiated by the wheels is represented by infinite line sources located 0.7 m above the surface. The attenuation is calculated by comparing the noise levels predicted at 7.5 m horizontal distance from the centre of the track and 1.3 m height above the railhead. The flat ground surface is modelled to represent grassland using $\Omega = 0.4$, $q = 1.581$ and $\sigma = 200000$ Ns/m^4 and an infinite layer thickness. In general, the impact of a modification to the porous properties of the ballast will be greatest for sound propagating a long distance over ballast. Therefore, the modifications will have the largest effect on noise emitted by the rail on the far side. The effect of the wheel on the near side to the receiver is limited since the noise emitted from this source is not reflected from the ballast surface in the direction of the receiver. A summary of the most significant predicted reductions in overall A-weighted sound levels for the far side rail is given in table 1. Assuming constant porosity, increases in flow resistivity to 0.5 and 10 kPa s m^{-2} correspond to reductions in mean stone size to by factors of 5 and 100 respectively.

Table 1. Summary of predicted increases in attenuation of overall A-weighted levels (dB) for a source at the far rail

Change	**Increase flow resistivity to 10 kPa s m^{-2}**	**Increase flow resistivity to 0.5 kPa s m^{-2}**	**Increase porosity to 0.5**	**Increase layer thickness to 1 m**
Predicted extra attenuation dB	1.0	1.4	0.6	0.6

5 CONCLUSIONS

The deployment of artificial porous materials for noise reduction in outdoor situations is receiving increasing attention. There are several possibilities for controlling outdoor ground effect associated with the presence of porous ground surfaces. These possibilities are associated with increased understanding of the factors that influence it. Flow resistivity has an important influence and so does near-surface layering. Recent work has identified the additional role of surface roughness. All of these properties are influenced by cultivation practices. The surface of a well-kept sports field will have less roughness than that of set-aside grassland. Sub-soiling or ploughing reduce the flow resistivity of a soil and increase its surface roughness. Changing the surface impedance of the ground will have the greatest effect on sound propagation where the noise source is close to the ground surface. The acoustical influence of railway ballast is predicted to be increased by changing stone size and porosity. Reduction in ground impedance alongside road or rail traffic by means of this knowledge can lead to significant noise reduction.

6 ACKNOWLEDGEMENTS

This work was supported in part by BBSRC (UK) Grant No. CTE 02649 and by a contract with M + P Raadgevende Ingenieurs bv, Aaslsmeer, the Netherlands. The author is grateful to Tim Waters-Fuller for the data obtained over grassland and cultivated soils and to Dr. H. A. Holties, of M + P Raadgevende Ingenieurs bv, for his contribution to the ballast attenuation study.

7 REFERENCES

[1] Attenborough, K. "Review of ground effects on outdoor propagation from continuous broadband sources", Applied Acoustics **24** 289-319 (1988)

[2] Price, M., Attenborough, K., and Heap, N. W., "Sound attenuation through trees: Measurements and models", J. Acoust. Soc. Am. **84** 1836 – 1844 (1986)

[3] Huisman, W. H. T., and Attenborough, K., "Reverberation and attenuation in a pine forest: models and measurements", J. Acoust. Soc. Am., 90 (5) 2664 – 2677 (1991)

[4] Attenborough, K. and Li, K.M. Ground effect for A-weighted noise in the presence of turbulence and weak refraction, J. Acoust. Soc. Am. **102** (2) 1013 – 1022 (1997)

[5] Attenborough, K. and Taherzadeh, S. Propagation of sound from a point source over a rough finite impedance surface J. Acoust. Soc. Am. **98** 1717 – 1722 (1995)

[6] Boulanger, P.M., Attenborough, K., Taherzadeh, S., Waters-Fuller, T and Li, K.M. "Ground effect over hard rough surfaces", J. Acoust. Soc. Am. **104** (3) 1474-1482 (1998)

[7] Watts, G., Transport Road Laboratory, U.K. private communication (1998)

[8] Aylor, D. "Noise reduction by vegetation and ground", J. Acoust. Soc. Am., **51** (1) 197 – 205 (1972)

[9] Attenborough, K., Waters-Fuller, T., and Li, K.M., "Acoustical properties of farmland", Proc. 3rd Agroacoustics Symposium, Iuka, Mississippi, USA, Sept. 1998 (in press)

[10] Taherzadeh, S. and Attenborough, K. "Deduction of ground impedance from measurements of excess attenuation spectra", J. Acoust. Soc. Am. (in press)

[11] "Calculation of Road Traffic Noise", Department of Transport Welsh Office, HMSO (1993)

[12] Hothersall, D.C. and Chandler-Wilde, S.N., "Prediction of L10 noise levels from road traffic: distance corrections for inhomogeneous ground cover", Proc. I. O. A., 10 (2) (1988)

[13] Boulanger, P., Waters-Fuller, T., Attenborough, K., and Li, K. M. "Models and measurements of propagation from a point source over mixed impedance ground", J. Acoust. Soc. Am. **102** (3) 1432 – 1442 (1997)

[14] Transportation Noise Reference Book, ed. P.M. Nelson, Butterworths, 1987

[15] D.C. Hothersall, S.N. Chandler-Wilde and M.N. Hajmirzae, "Efficiency of single noise barriers", J. Sound Vib., **146**, p. 303 (1991)

[16] K. Attenborough and C. Howorth, "Models for the acoustic characteristics of porous road surfaces", Proceedings Int. Tire/Road noise conf. 1990, Gothenburg, p. 177 (1990)

[17] "Richtlijnen voor de bovenbouw (ballastbed)", 1 April 1992

[18] CEN TC154 SC4 Ad Hoc group Aggregates for Railway Ballast, "Draft technical specifications for railway ballast", 3 January 1997.

[19] W.O. Smith, Physics 3, (1932)

[20] Ministerie VROM, Directie Geluid en Verkeer, Reken- en meetvoorschrift Railverkeerslawaai '96, 20 November 1996

S682/004/99

Sound power flux measurements in exhaust systems

P O A L DAVIES and **K R HOLLAND**
Institute of Sound and Vibration Research, Southampton University, UK
D VAN DER WALT
Bosal Afrika, South Africa

1 INTRODUCTION

Measurements of orifice noise emissions from intake and exhaust during controlled engine acceleration and run down on an appropriate test bed normally form one essential part of component design refinement during prototype vehicle development [1] They are used to indicate those spectral regions in the exhaust noise signature where further reductions in the level are necessary or improvements in the sound quality are desirable. Engine noise spectra exhibit order noise, that is a sequence of tonal components associated with the cyclicly pulsating flow through the valves that are thus harmonically related to the firing frequency. They also include broad band components from flow noise that normally provides the major spectral contributions above 300Hz [1, 2]. This arises from turbulence and other flow-induced aeroacoustic sources associated with shear layers, flow separations, wakes and vortex generation occurring at expansions, contractions, bends and other discontinuities distributed along the system. Furthermore [3, 4] vortex excitation of expansion chambers, coupled with acoustic feedback, can produce significant resonant amplification of order noise [1, 4] at relevant coincident frequencies.

It is widely acknowledged that noise emission measurements alone [1] cannot normally provide specific indications as to where and what detailed changes to system geometry or layout will lead directly to the desired refinement or acoustic performance. Relevant interpretations of the data depend on physical insight and experience, assisted in a rather limited way by information in the rapidly expanding literature [1] describing sound propagation in exhaust systems. Currently, its potential value is restricted to order noise assessment, since this material does not include enough information [1–4] to provide quantitative estimates of the position and acoustic characteristics of any flow noise sources distributed along the system.

An alternative approach is to obtain direct measurements of the acoustic behaviour at appropriate positions along the exhaust system by adopting experimental techniques using flush wall mounted pressure transducers. Such methods have been available since the development of digital signal acquisition and processing facilities in the early 1970's [5, 6]. In particular, the aim would be to identify and quantify the flow noise sources by measuring, with a sequence of microphone pairs, the distribution of acoustic power flux at appropriate positions along the system. However, it turns out that measuring acoustic properties, such as transfer functions and acoustic power flux, in ducts and pipes, in the presence of significant flow, is fraught with difficulty, mainly since boundary layer and other flow-induced pressure fluctuations at or near the wall can give rise to very poor signal to noise ratios. In what follows, techniques are described that are intended to significantly reduce signal-to-noise ratio problems associated with wall mounted pressure transducers and flow generated noise.

The result of two sets of experiments are reported here. One set, performed in the laboratory, was concerned mainly with the development and evaluation of the proposed new technique by comparison first with the results obtained by the more conventional white noise excitation and second by comparisons with those obtained with the well validated predictive software package APEX. The other experiments were carried out in an industrial development facility for non-stationary conditions by the third author. They include simulated acceleration tests on a cold flow test bench with sawtooth excitation combined with similar measurement made during a sequence of engine acceleration tests. The observed attenuation performance of the test expansion chamber, at the first four engine orders (2E, 4E, etc) was again compared in each case with the corresponding predictions obtained with a suitably modified version of the software package.

2 MEASUREMENT METHOD

The two-microphone method is a well established [5, 6] and proven technique for estimating the acoustic properties of, and the acoustic quantities within, a duct system; pairs of microphones are placed axially along each section of duct of interest, and the auto- and cross-spectra determined from the microphone pressure-time histories may then be processed [4–6] to yield estimates of pressure spectra, forward and backward wave components, system transfer functions or sound power flux in the corresponding pipe or duct.

2.1 Determination of Wave Components

Consider two microphones spaced a distance l apart. The fluctuating acoustic pressure at any point and frequency may be expressed as the sum of forward (in the direction of flow) and backward propagating wave components having complex spectral amplitudes $\hat{p}^+$ and $\hat{p}^-$ respectively, thus,

$$\hat{p}(x, \omega) = \hat{p}^+ e^{-j\hat{k}^+ x} + \hat{p}^- e^{j\hat{k}^- x}, \tag{1}$$

where, to first order in Mach number ($M = u_0 / c_0$)

$$\hat{k}^+ = \frac{k + \alpha(1 - j)}{(1 + M)} \text{ and } \hat{k}^- = \frac{k + \alpha(1 - j)}{(1 - M)}, \tag{2a, b}$$

are the forward and backward complex wavenumbers respectively, $k = w/c_0$, a is the viscothermal attenuation coefficient. Arbitrarily assigning $x = 0$ to be the position of microphone #1, the pressures at the two microphones may be written

$$\hat{p}_1(\omega) = \hat{p}^+ + \hat{p}^-, \text{ and } \hat{p}_2(\omega) = \hat{p}^+ e^{-j\hat{k}^+ l} + \hat{p}^- e^{j\hat{k}^- l}. \qquad (3a, b)$$

Let $\hat{G}_{12}$ be the cross spectrum between the pressures at microphones #1 and #2 estimated from the time histories for the two microphones, and let G_{11} be the autospectrum of the pressure at microphone #1. Writing the complex ratio of the backward to forward pressure wave amplitudes as a complex pressure reflection coefficient ($\hat{R}$) at the position $x = 0$ yields

$$\hat{R} = -\left\{\frac{\hat{G}_{12} - G_{11} e^{-j\hat{k}^+ l}}{\hat{G}_{12} - G_{11} e^{j\hat{k}^- l}}\right\}, \qquad (4)$$

from which,

$$|\hat{p}^+| = \frac{(G_{11})^{1/2}}{|1 + \hat{R}|} \text{ and } \hat{p}^- = \hat{R}\hat{p}^+. \qquad (5a, b)$$

2.2 Determination of Sound Power Flux

The sound intensity and hence sound power flux in a section of the duct may be calculated from the forward and backward pressure wave components estimated using equations (5) above [7], thus

$$I_T = I^+ - I^- = \frac{1}{\rho_0 c_0}\left(|\hat{p}^+|^2 (1 + M)^2 - |\hat{p}^-|^2 (1 - M)^2\right). \qquad (6a, b)$$

2.3 Microphone Calibration

The acoustic field within a duct system is necessarily highly reactive at the low Helmholtz numbers *(ka)* of interest, thus the phase, and hence the imaginary part, of the crosspectra used in the estimation of power flux is very close to zero over much of the frequency band. Accurate, in-situ calibration of the microphones is therefore crucial to the success or otherwise of the subsequent measurement methods. The calibration method chosen involves physically interchanging two of the microphones thus: the two microphones are placed in position and the transfer function between their outputs is measured ($\hat{H}_A$). The two microphones are then interchanged, including all signal path elements (recorder channel, amplifier etc.) and the measurement repeated ($\hat{H}_B$). The required (calibration independent) transfer function of interest is then $\hat{H} = (\hat{H}_A \hat{H}_B)^{1/2}$, and the relative microphone calibration is $(\hat{H}_A / \hat{H}_B)^{1/2}$. This method is repeated for each of the microphones in turn yielding relative microphone calibration data; the (approximate) absolute calibration of one microphone only is then required if absolute sound pressure levels are important.

3 DATA ACQUISITION AND ANALYSIS METHOD

The output signals from the 5 microphones shown in Figure 1 are recorded onto a multi-track digital tape recorder which is then played back into a multi-channel computer data acquisition system and stored as 16-bit data. This method gives freedom for experimentation with various data analysis methods and is therefore preferred, in this case, over the use of a dedicated multi-channel analysis package. All analysis was carried out using custom written code on a 200MHz RISC computer.

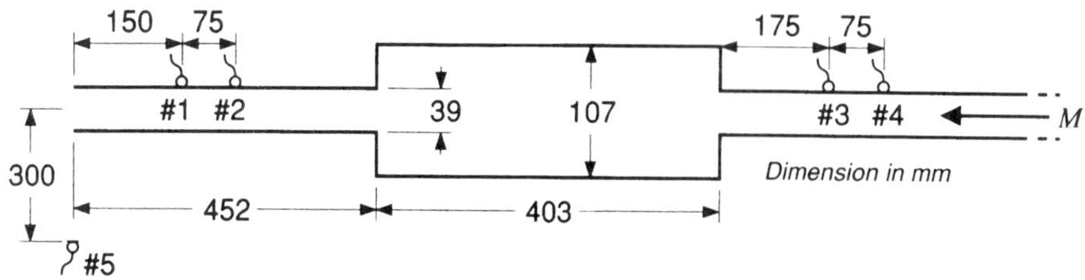

Figure 1 Geometry of Duct System

3.1 Random Excitation

Experiments were carried out using a white noise signal as input to the loudspeaker. This form of acoustic excitation is commonplace in both static and flow acoustic studies and is the automatic 'first choice' broadband test signal. Figure 2a shows the H_1 transfer function ($\hat{p}_2 / \hat{p}_1$) measured between microphones #1 and #2 of the duct system using white noise excitation with a flow Mach number $M = 0.1$; Figure 2b shows the coherence function for the measurement. The sound pressure level of the acoustic signal at microphones #1 and #2 was 131dB SPL which was greater than 10dB above that of the boundary layer noise.

As the microphones are mounted close to a simple open pipe end, one would expect the transfer function between them to be a relatively simple function of frequency. This is not the case in Figure 2a, and inspection of the coherence (Figure 2b) reveals only limited bands of frequencies over which the measured result can be considered valid. Clearly, although the rms level of the acoustic excitation is much larger than that of the boundary layer noise, the wide variations in acoustic level with frequency at the exit of a resonant expansion chamber yield inadequate signal-to-noise ratios over much of the bandwidth. As the acoustic signal and the boundary layer noise are both random, separation of the two is difficult. Figure 2c compares the autospectrum of the output from microphone #1 observed with flow and acoustic excitation and with flow alone; a comparison between these and Figure 3c (see Section 3.2) also shows that the dips in acoustic signal level were buried in the noise floor. Raising the acoustic excitation level to help overcome the signal-to-noise ratio problem is impractical for three reasons: the loudspeaker may overload: the microphones may overload: the propagation of acoustic waves will become nonlinear. Results such as that shown in Figure 2 are unacceptable for acoustic power flux estimation; a more precise method has been devised which utilises a slow sine-sweep signal.

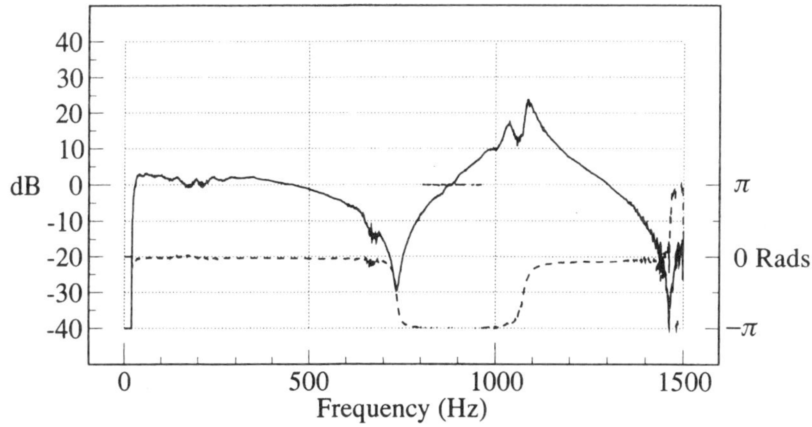

Figure 2a Modulus (solid) and Phase (dashed) of Transfer Function $\hat{H}_{12}$ Measured using Random Excitation with Mach Number M = 0·1

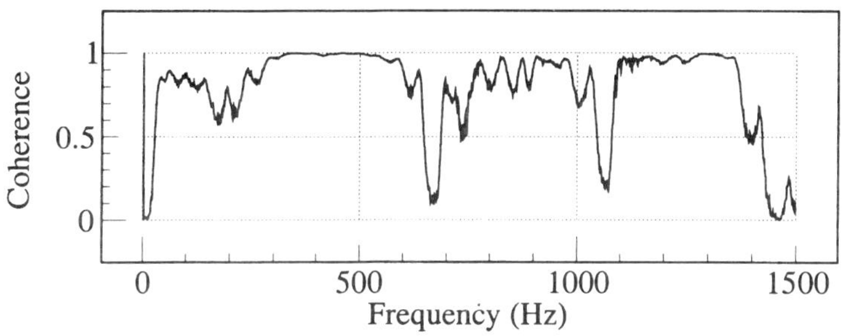

Figure 2b Coherence Function for Measurement in 2a

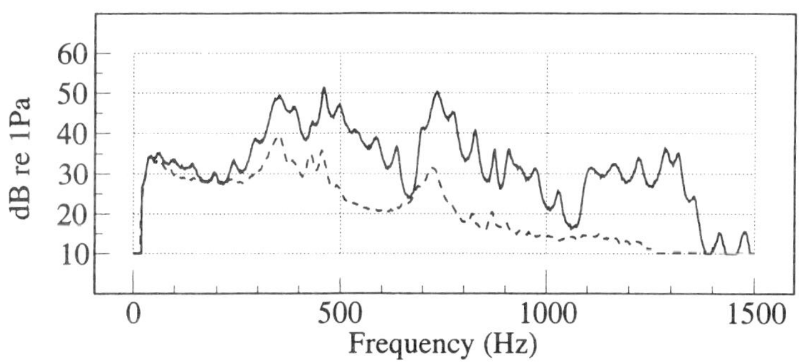

Figure 2c Autospectrum of Output of Microphone #1 for Measurement in 2a with Acoustic Excitation (solid) and without (dashed)

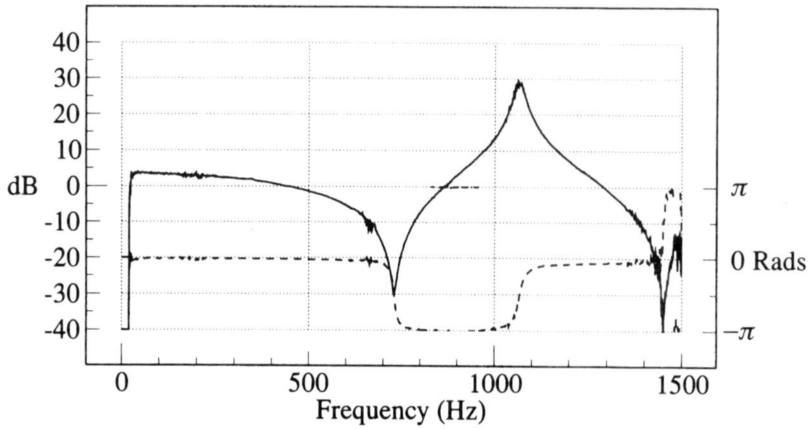

Figure 3a Modulus (solid) and Phase (dashed) of Transfer Function $\hat{H}_{12}$ Measured using Swept-Sine Excitation with Selective Averaging: Mach Number $M = 0{\cdot}1$.

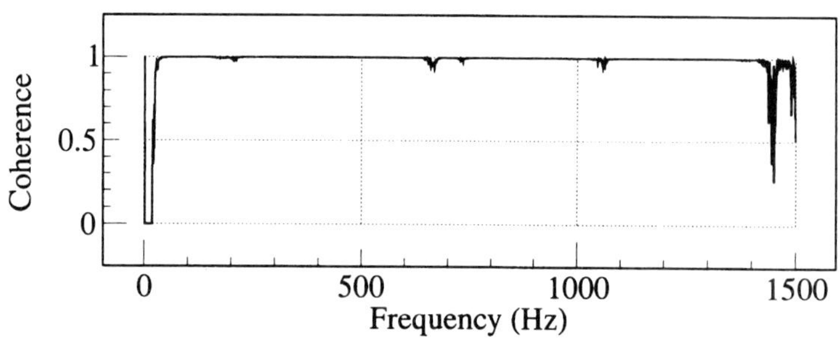

Figure 3b Coherence Function for Measurement in 3a

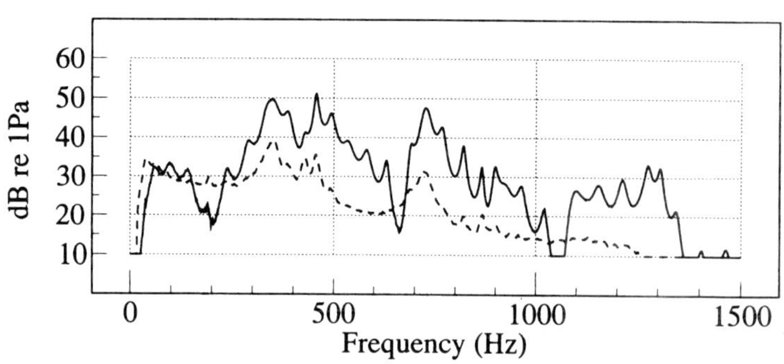

Figure 3c Autospectrum of Output of Microphone #1 for Measurement in 3a with Acoustic Excitation (solid) and without (dashed)

3.2 Sine-Sweep Excitation

In an attempt to overcome the signal-to-noise ratio problems encountered using random noise excitation, a slow sine-sweep signal was used to excite the duct system. This signal was generated digitally and recorded onto one channel of the digital tape; the remaining channels were then used to record the pressure signals from the microphones. Thus the excitation signal and the resulting pressure signals were always synchronous in time. Furthermore, it was then relatively easy to tailor the digitally generated signal so that the sweep rate varied with time; setting a slower sweep rate at low frequencies, where the signal-to-noise ratio is poorest, resulted in improvements in the results at low frequencies. It is also possible to tailor the amplitude of the signal to improve the signal-to-noise ratio at some frequencies; this was not attempted for the measurements described herein. Because the frequency of the sweep is known at any particular point in time, it is possible to exclude from the averaging process all frequencies except those within a narrow band around the known centre frequency. Figure 3a shows the results of carrying out analysis of this nature on the microphone output signals with excitation in the form of a 10 minute swept-sine signal and a flow Mach number $M = 0.1$. The sound pressure level of' the acoustic excitation varied with frequency between 100dB and 140dB at microphones #1 and #2; the overall time-averaged level being 131dB; Figure 3b shows the coherence function for this measurement, and Figure 3c, the autospectrum of the output of microphone #1 with and without acoustic excitation.

A Comparison between Figures 2, using random excitation and conventional analysis, and Figures 3, using swept-sine excitation and selective averaging, shows that it is possible to separate out the wanted acoustic signal from the random boundary layer noise using swept-sine techniques, even when the signal-to-noise ratio is much less than unity.

4 LABORATORY MEASUREMENTS COMPARED WITH PREDICTIONS

Computer codes exist for the prediction of the acoustic properties of ducts with flow. The predictions of one such code (APEX, see [8, 9]), written by the first author, are presented below for comparison with the measured results. Good agreement between the predictions and measurements is assumed to validate both the code and the measurement technique.

4.1 Acoustic Transfer Functions

Figure 4 shows a comparison between the predicted and measured (using swept-sine excitation and selective averaging) transfer functions $\hat{H}_{13}$ between microphones #1 and #3 with a flow Mach number $M = 0.1$.

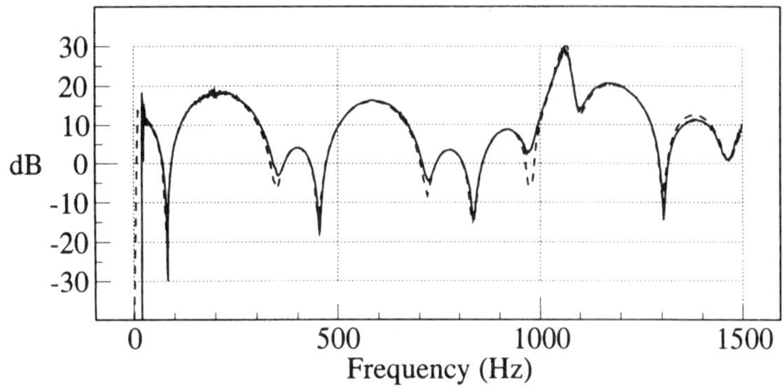

Figure 4 Measured (solid) vs Predicted (dashed) Transfer Function $\hat{H}_{13}$ with Mach Number M = 0·1

4.2 Acoustic Power Flux

The measured transfer functions were used to estimate the forward and backward pressure wave components using Equation (5), from which the net acoustic power flux may be estimated according to Equation (6). Figures 5a and 5b show comparisons between the predicted and measured net acoustic power flux in the outlet pipe and also for the radiated power with a flow Mach number $M = 0.1$. For clarity and ease of comparison, the results and predictions are all normalised to the forward (with flow) acoustic power flux in the inlet pipe.

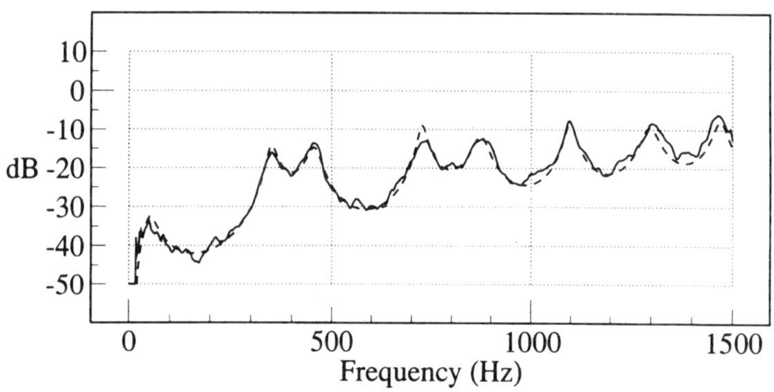

Figure 5b Measured (solid) vs Predicted (dashed) Normalised Radiated Acoustic Power with Mach Number M = 0.1

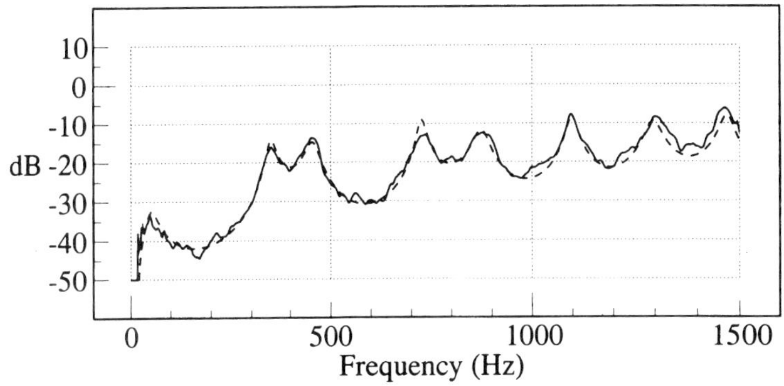

Figure 5b Measured (solid) vs Predicted (dashed) Normalised Radiated Acoustic Power with Mach Number M = 0.1

5 MEASUREMENTS IN AN INDUSTRIAL ENVIRONMENT

As noted in the introduction, the schedule for these measurements included both those carried out during bench tests with corresponding sequences made on an engine dyno. rig. The bench test simulations were primarily intended to confirm the operational integrity and suitability of the Brüel & Kjær data capture equipment, the Kistler 701a water-cooled pressure transducers and so on. Also to investigate and evaluate proposed experimental and data processing procedures and finally to produce baseline information describing the acoustic behaviour of the various silencer elements. The excitation and flow were controlled to simulate engine acceleration conditions. The resulting order measurements were compared with corresponding predictions obtained with appropriately extended APEX software. Close agreement was found between measurements and predictions (within ±1dB) for the first three orders and also for the 4th order below 600Hz. The slightly larger differences found above this frequency were associated with exceptionally low signal-to-noise ratios.

5.1 Attenuation Measurements During Engine Acceleration

The test elements were then incorporated into an engine silencer system and their attenuation performance, defined as the change in incident wave amplitude on each side of the element, was evaluated from the four transducer measurements. The mass flow, engine rotational frequency and relevant exhaust temperatures were recorded to provide the additional data needed to perform a corresponding prediction of the attenuation. The rotational frequency was also required for the order tracking analysis procedures that had been verified during the test bench experiments. Representative measurements are compared with predictions in Figure 6. Again, the discrepancies illustrated here could all be traced to excessively low signal-to-noise ratios. It turns out however, that consistent estimates were not obtained for the corresponding acoustic energy flux. Clearly further development of the experimental procedures is necessary before such estimates can be produced.

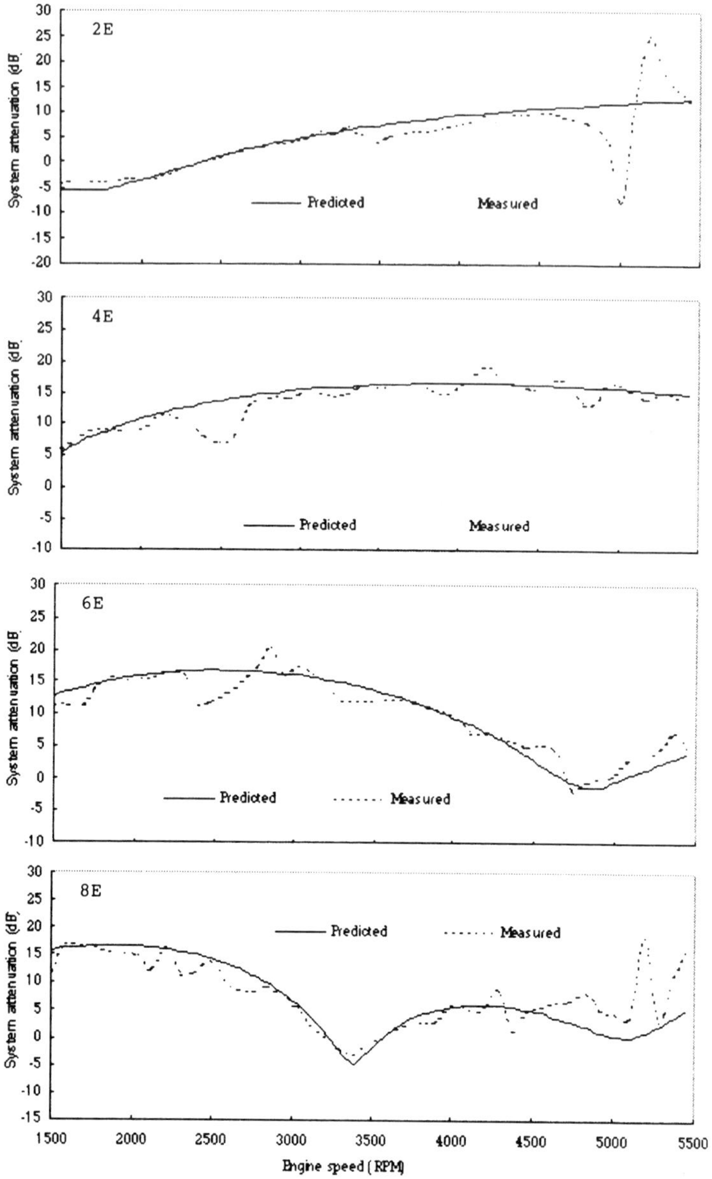

Figure 6 Measured vs Predicted System Attenuation – Engine Acceleration Test

6 DISCUSSION AND CONCLUSIONS

The driving force behind this investigation is the study of flow-induced sources within flow duct systems. It turns out [1–4] that sufficiently precise measurement of power flux are required on either side of the suspected source sites, to identify and quantify such sources in real systems. The results have shown that careful calibration of the microphones is essential and that suitable experimental and data processing procedures are also essential to secure realistic measurements of power flux, when the signal-to-noise ratios are low. For such conditions, it was shown that slow swept sine excitation with synchronised selective signal averaging proved superior to conventional white noise excitation.

One might expect that similar procedures could be transferred to engine acceleration tests. This proved to be true for the simulations during cold flow bench tests but it has not so far been possible to obtain sufficiently precise measurements of power flux with an engine accelerating under load. Why this should be so is not fully understood and investigations to rectify this are in progress.

One useful outcome of the more precise acoustic measurements that have been obtained in both cases is a more exact and realistic validation of the predictive software package used for comparison with the measurements. The agreement seems sufficiently close in general for one to assume that the converse is also true, and thus applies to the assessment of the effectiveness of the experimental procedures as well.

7 ACKNOWLEDGEMENTS

The laboratory experiments form part of the studies of sources of flow-induced pressure pulsations in pipe systems transporting compressible fluids supported by the Engineering and Physical Science Research Council. Thanks are due to Bosal Afrika for their financial support of the investigations by the third author.

8 REFERENCES

[1] P. O. A. L. Davies and K. R. Holland 1999. *Journal of Sound and Vibration*, **223**, 425–444. I. C. engine intake and exhaust noise assessment.

[2] P. O. A. L. Davies 1996 *Journal of Sound and Vibration*, **190**, 345–362. Aeroacoustics and time varying systems.

[3] T. D. Mast and A. D. Pierce 1995 *Journal of the Acoustic Society of America*, **97**, 163–172. Describing function theory for flow excited resonators.

[4] P.O. A. L. Davies 1981 *Journal of Sound and Vibration*, **77**, 191–209. Flow acoustic coupling in pipes.

[5] J. E. Temple 1990 MSc Thesis *University of Southampton.* An investigation into the source region characteristics of internal combustion engine exhaust systems.

[6] J. Y. Chung and D. A. Blaser 1980 *Journal of the Acoustic Society of America,* **68**, 1570–1577. Transfer function method of measuring acoustic intensity in a duct system with flow.

[7] C. L. Morfey 1971 *Journal of Sound and Vibration* **14**, 37–55. Sound generation and transmission in ducts with flow.

[8] P. O. A. L. Davies, M. F. Harrison and H. J. Collins 1997 *Journal of Sound and Vibration,* **200**(2), 195–225. Acoustic modelling of multiple path silencers with experimental validations.

[9] P. O. A. L. Davies and M. F. Harrison 1997 *Journal of Sound and Vibration*, **202**(2), 249–274. Predictive acoustic modelling applied to the control of intake/exhaust noise of internal combustion engines.

S682/005/99

Theoretical prediction of ground vibrations generated by road vehicles

V KRYLOV
Department of Civil and Structural Engineering, Nottingham Trent University, UK

Ground-borne vibrations represent one of the major adverse environmental impacts of road vehicles, especially of heavy lorries. In the present paper, ground elastic waves generated by road vehicles are investigated theoretically. Two main generation mechanisms are considered. The first one is associated with vehicles travelling on rough or bumpy road surfaces, in particular over road humps and speed cushions installed by local authorities at some sensitive road locations as a simple method of traffic calming. The second mechanism of generation is associated with acceleration and braking of road vehicles. General analytical results are illustrated by numerical examples and are compared with the existing experiments.

1 INTRODUCTION

Ground-borne vibrations represent one of the major adverse environmental impacts of road vehicles, especially of heavy lorries. Generally, one can distinguish two main mechanisms of generating ground vibrations by road vehicles (Watts 1984).

The first one is associated with vehicles travelling on rough or bumpy road surfaces, in particular over road humps and speed cushions installed by local authorities at some sensitive road locations as a simple method of traffic calming. As the practice show, installation of road humps and speed cushions results in significant reduction of the severity and number of road accidents as a result of decrease in speeds at which drivers choose to travel. However, the recent experimental investigations carried out by the UK Transport Research Laboratory (TRL) have demonstrated that such a gain is achieved partly in expense of the increased traffic noise and ground-borne vibration (Watts 1998). It has been shown during these investigations that speed control cushions and road humps can produce perceptible levels of ground vibration which depend on vehicle type and hump and cushion design.

The second mechanism of generation is associated with acceleration and braking of road vehicles. It is essential in the places where vehicles stop, start or change direction of their movement, e.g., before traffic lights or in supermarket yards.

Theoretical investigations of ground vibrations generated by road traffic have been carried out by several authors. A number of works has been done for vehicles travelling on statistically rough surfaces of rather good quality (Taniguchi et al. 1979, Le Houedec et al. 1982, Hunt 1991, Hanazato et al. 1991) and for accelerating and braking vehicles (Krylov 1995a-c, 1996a,b). However, vibrations caused by vehicles travelling over single obstacles were analysed only very briefly (Krylov 1995a,c) and no calculations directly relevant to traffic calming road humps and speed cushions were carried out. In the present paper, we describe a more comprehensive theory of ground elastic waves generated by road vehicles.

In the case of ground vibrations generated on bumpy roads, a vehicle is modelled as a mechanical system having four degrees of freedom associated with its main low-frequency resonances. Solving the system of the corresponding dynamic equations, the elastic Lame equation describing ground motion, and taking into account the boundary conditions on the ground surface, which incorporate dynamic forces caused by the interaction of a moving vehicle with a hump, results in the analytical expressions for generated ground vibration spectra. It is shown that amplitudes and spectra of generated ground vibrations depend on shapes and sizes of the road humps, on vehicle speeds and loads, and on ground elastic properties. In particular, the dependence on vehicle speeds demonstrates the initial increase of amplitudes as speeds increase and then their monotonous decrease according to the inverse proportionality law.

The problem of generating ground vibrations during acceleration and braking is considered for vehicles accelerating (decelerating) with a constant acceleration from rest to a constant speed, or braking to a stop from a constant speed. According to the low-frequency approximation considered, an accelerating or braking vehicle is modelled as a horizontal traction force proportional to its mass and acceleration and moving along with the vehicle. Frequency spectra of the generated surface waves are described as functions of acceleration and final (initial) speed of the vehicle. In particular, it is shown that for medium and upper bands of the spectra the generated ground vibration level is determined mainly by acceleration and is almost independent of the initial (final) vehicle speed. For low-frequency spectral bands, oscillations of ground vibration amplitudes versus both acceleration and initial speed may take place. These oscillations may be responsible for large statistical deviations of experimentally observed ground vibration levels.

2 GROUND VIBRATIONS CAUSED BY ROAD HUMPS AND CUSHIONS

2.1 Vehicle interaction with road humps

In what follows we give a very brief description of the mechanism of vehicle interaction with road humps and speed cushions (Fig. 1) necessary for understanding the process of generating ground vibrations.

Typical mechanical model of a road vehicle travelling on uneven roads possesses four degrees of freedom corresponding to four main resonance frequencies of low-frequency vibrations related to body bounce and pitch, or to front- and rear-axle hops (Hunt 1991, Wong 1993). Frequencies of body bounce and pitch resonances are normally very low (in the range of 1-3 Hz). Axle-hop resonance frequencies are essentially larger (from 8 Hz to 12 Hz) and are therefore more important from the point of view of generating ground vibrations (we remind the reader that ground wave generation efficiency is higher at higher frequencies).

Keeping this in mind, we use the simplified model of a vehicle, considering its carriage as immobile in vertical direction and taking into account only axle vibrations. This model consists of two identical vibrating systems, each having one degree of freedom and comprising an axle mass m and two springs with constants K_1 and K_2 modelling respectively the elasticity of tyre and suspension (Fig. 2). Axles are separated from each other by the distance L (wheel base). We also assume that a hump or cushion cross-section in the plane $y = 0$ is described by the function $z_1 = f(x)$ (see Fig. 1).

According to the model considered, the equation describing vertical displacements of each axle versus its static position z_2 has the form

$$m\partial^2 z_2/\partial t^2 + Q\partial z_2/\partial t + Kz_2 = K_1 z_1(vt), \tag{2.1}$$

where $K = K_1 + K_2$ is a combined elasticity of tyre and suspension, and Q is a total damping coefficient.

Assuming that hump area is small as compared with wavelengths of generated vibrations and its centre is located at $x = 0$ and $y = 0$, the related normal stress T_{zz} applied to the ground can be written in the form

$$T_{zz}(\rho,t) = K_1[(z_2(t) - Z_1(t) + z_2(t-L/v) - Z_1(t-L/v)]\delta(x)\delta(y), \tag{2.2}$$

where $Z_1(t) = z_1(vt)$ and $Z_1(t-L/v) = z_1(vt-L)$ are the input functions for the front and rear axles respectively, $\rho = \{x, y\}$ is the surface radius-vector, and $\delta(z)$ is Dirac's delta-function.

Solving eqn (2.1) by Fourier method, one can derive the following expression for the Fourier transform $z_2(\omega)$ of a front wheel axle vertical displacement $z_2(t)$:

$$z_2(\omega) = \frac{\omega_1^2 Z_1(\omega)}{\sqrt{(\omega_0^2 - \omega^2)^2 + (2\omega\alpha)^2}} exp[-i\, tan^{-1}(\frac{2\omega\alpha}{\omega_0^2 - \omega^2})], \tag{2.3}$$

where $\omega_0 = (K/m)^{1/2}$ is the hop resonance frequency, $\omega_1 = (K_1/m)^{1/2}$ is the tyre "jumping" resonant frequency, $\alpha = Q/2m$ is a normalised damping coefficient, and $Z_1(\omega)$ is the Fourier spectrum corresponding to the hump profile. The Fourier transform for a rear wheel axle vertical displacement $z_2(t-L/v)$ differs from (2.3) only in phase shift $\omega L/v$ and is omitted here for shortness. The Fourier transform of the vertical force applied from the vehicle to the ground, $T_{zz}(\rho,\omega)$, is easily obtained from (2.2) via replacing $z_2(t)$, $Z_1(t)$, $z_2(t-L/v)$, and $Z_1(t-L/v)]$ by their Fourier spectra:

$$T_{zz}(\rho,\omega) = K_1[(z_2(\omega) - Z_1(\omega) + z_2(\omega)e^{i\omega L/v} - Z_1(\omega)e^{i\omega L/v}]\delta(x)\delta(y). \tag{2.4}$$

2.2 Calculation of generated elastic waves

The ground vibration field generated by vehicles in an elastic half space, which we assume to be homogeneous and isotropic, should satisfy the elastic Lame′ equation

$$(\lambda+2\mu)\ grad\ div\ \boldsymbol{u} - \mu\ rot\ rot\ \boldsymbol{u} - \rho_0 \partial^2 \boldsymbol{u}/\partial t^2 = 0\ , \qquad (2.5)$$

and the boundary conditions on the ground surface taking into account the vertical force (2.2) resulting from the interaction of a vehicle with an uneven surface. Here $\boldsymbol{u}$ is the particle displacement vector with the components u_i ; λ and μ are the elastic Lame′ constants; and ρ_0 is the ground mass density.

The ground surface geometry is rather complex due to the presence of road humps and cushions. However, since the characteristic height h of road humps and cushions is very small in comparison with characteristic wavelengths λ_v of generated vibrations, it can be simplified by using the method of projection of the actual boundary conditions onto the flat surface $z = 0$. One can show that the projected boundary conditions differ from the real ones only by the additional small terms proportional to $h/\lambda_v << 1$. In further consideration we will neglect these terms and write the boundary conditions at $z = 0$ in the form:

$$\begin{aligned} \sigma_{xz} &= 2\mu u_{xz} = 0 \\ \sigma_{yz} &= 2\mu u_{yz} = 0, \\ \sigma_{zz} &= \lambda u_{nn} + 2\mu u_{zz} = -T_{zz}(\rho,t). \end{aligned} \qquad (2.6)$$

Here $u_{ij} = (1/2)(\partial u_i/\partial x_j + \partial u_j/\partial x_i)$, where $i,j = 1,2,3$, are the components of the linearised deformation tensor; and σ_{ij} are the components of the elastic stress tensor. Without loss of generality, we limit our calculation of frequency spectra to the vertical component of ground vibration velocity $v_z = du_z/dt$ which is usually measured during experimental observations.

Solving the boundary value problem (2.2)-(2.6) by the Green's function method (see, e.g., Ewing et al. 1957; Graff 1975) and taking into account only generated Rayleigh waves, one can derive the following expression for the vertical component of the surface vibration velocity spectrum:

$$v_z(\rho,\omega) = \left(\frac{2\pi}{k_R\rho}\right)^{1/2} \frac{(-i\omega)k_R k_t^2 v_l}{2\pi\mu F'(k_R)} T_{zz}(\omega) e^{-k_R\gamma\rho} e^{ik_R\rho - i3\pi/4} . \qquad (2.7)$$

Here $T_{zz}(\omega) = T_{zz}(0,\omega)$ (see eqn (2.4)); $\rho = \rho(x,y)$ is the distance to the observation point; $k_R = \omega/c_R$ is the Rayleigh wave number, where c_R is Rayleigh wave velocity; $v_{l,t} = (k_R^2 - k_{l,t}^2)^{1/2}$ are nonspecified expressions, where $k_{l,t} = \omega/c_{l,t}$ are the wavenumbers of bulk longitudinal and shear acoustic waves, c_l and c_t are their phase velocities; $F'(k_R)$ is the derivative $dF(k)/dk$ of the Rayleigh determinant $F(k) = (2k^2 - k_t^2)^2 - 4k^2 v_l v_t$ taken at $k = k_R$.

In writing (2.7) we have taken account of attenuation of generated ground vibrations in the ground by replacing the wavenumber of a Rayleigh wave in an ideal elastic medium $k_R = \omega/c_R$ by the complex wavenumber $k_R' = k_R(1 + i\gamma) = (\omega/c_R)(1+i\gamma)$. Here $\gamma << 1$ is a positive constant which describes the linear dependence of a Rayleigh wave attenuation coefficient on frequency ω. For different types of ground γ are in the range from 0.01 to 0.2 (Gutovski et al. 1976). In what follows we will be interested only in amplitudes of ground vibrations $V_z(\rho, \omega) = |v_z(\rho,\omega)|$, ignoring the phase information.

2.3 Numerical calculations and discussion

Numerical calculations of ground vibrations generated by vehicles crossing humps and cushions have been carried out according to eqn (2.7) for vehicles travelling at different speeds over road humps and cushions of different profiles and dimensions. Vehicle parameters were: $\omega_1 = 65\ rad/s$, $\omega_0 = 70\ rad/s$, $\alpha = 15$, $L = 6\ m$, and $m = 50\ kg$. Ground parameters were: Rayleigh wave velocity $c_R = 125\ m/s$, Poison ratio $\sigma = 0.25$, ground mass density $\rho_0 = 2000\ kg/m^3$, ground attenuation constant $\gamma = 0.05$. Observation distance was $\rho = 30\ m$.

Obviously, if the spectrum $Z_l(\omega)$ has components around the axle hop resonance frequency ω_0, then the axle vibrations are effectively excited and noticeable generation of ground vibrations takes place. Analysis and numerical calculations show that amplitudes and spectra of generated ground vibrations depend on hump and cushion profile, their characteristic gradient (ratio h/l), vehicle speed v and load, and elastic parameters of the ground.

In particular, the effect of hump shape on ground vibration spectra generated by vehicles travelling at speed $v = 5\ m/s$ is demonstrated in Fig. 3 for three hump types having the same characteristic height h and roughly the same length l, but different geometrical shapes: a Gaussian (bell-shaped) hump, $z_l(x) = h\ exp(-x^2/l^2)$ (curve V1z), cosine shaped hump, $z_l(x) = h\ cos(\pi x/l)$ (curve V2z) and rectangular hump, $z_l(x) = h\ [H(x+l/2) - H(x-l/2)]$ (curve V3z), where $H(x)$ is a step function. Calculations were carried out for humps with the following typical parameters: $h = 0.074\ m$, and $l = 0.9\ m$. One can see that the worst hump shape which causes the highest level of generated vibrations is a rectangular shape, and the best - a Gaussian one. The replacement of typical cosine-shaped humps by Gaussian ones would cause approximately 20 dB reduction in the averaged level of generated ground vibrations.

Calculations for cosine-shaped road humps, including the ones with typical parameters: $h = 0.074\ m$, $l = 0.9\ m$; and $h = 0.064\ m$, $l = 3.7\ m$, show that the integral level of generated ground vibrations grows roughly in direct proportion to h/l. This agrees well with the experimental results of TRL for two round-shaped profiles of the above mentioned dimensions (Watts 1998).

Dependence of generated ground vibrations on vehicle speed v is quite complex for each particular spectral component. Calculated results show that for small and medium speeds v ground vibration amplitudes grow with oscillations (depending on hump length l and wheel base L) and reach a maximum at certain v. Then, for larger values of v, the level of generated vibrations decays inversely proportionally to v.

Calculation of the integral level of generated ground vibrations carried out for the same three types of humps, a Gaussian (bell-shaped) hump (curve Iv1z), cosine shaped hump (curve Iv2z) and rectangular hump (curve Iv3z), show that it depends on v in a more simple way (Fig. 4). Note that for Gaussian and cosine shapes the dependence is roughly linear (with different coefficients of proportionality) for typical values of v at which drivers chose to travel in traffic calming areas (from 0 to 7 m/s). Again, the initial average growth of generated ground vibrations with the increase of vehicle speed agrees with the TRL experimental results (Watts 1998), although no detailed comparison is possible because of the insufficient number of vehicle speeds (four) used in the experimental observations. Also, one can clearly see that the integral level of generated ground vibrations is the lowest for a hump of Gaussian shape.

Note that heavy loads result in the increase of static stresses applied to the vehicle suspension which, due to nonlinearity of material stress-strain relations, will change the stiffness of an equivalent spring and, hence, modify the axle hop resonance frequency ω_0. The directions of these modifications are different for steel and air suspensions: for the first case the frequency ω_0 decreases, whereas for the latter case it increases. According to the above theory and in agreement with the experiments (Watts 1998), these may result in a small decrease in generated vibrations for vehicles with steel suspension and in their small increase for vehicles with air suspension.

The presence of the ground shear modulus μ in the denominator of eqn (2.7) implies that for the softer ground (smaller values of μ) the generated ground vibrations are more intensive.

3 GROUND VIBRATIONS CAUSED BY ACCELERATION AND BRAKING

3.1 Traction forces applied to the ground

In this section we consider generation of ground vibrations by vehicles accelerating at constant acceleration a from rest to a constant speed v, or braking with constant acceleration $-a$ from v to a complete stop. The assumption of constant acceleration (deceleration) is in agreement with general requirements for vehicle performance, e.g., in the process of braking. According to these requirements (Newcomb et al 1967), deceleration produced during braking should ideally be uniform throughout an application.

It should be emphasised that, in contrast to the normal load forces considered in the previous section, the main mechanism of ground vibration generation by accelerating or decelerating vehicles is the action of horizontal traction forces which are applied from tyres to the ground only during the time of accelerating or braking (the contribution of normal load forces is relatively small for speeds and accelerations typical for road traffic). Note that the horizontal interaction forces, acting during deceleration between vehicle and road through the tyre-road interface, are limited by friction (adhesion) between tyre and road. If the product of deceleration and vehicle mass exceeds this friction, sliding or skidding will occur.

Road braking statistics show that in the majority of cases decelerations range from *0.1 to 0.2g*, i.e. from about *1 to 2* m/s^2. However, in emergency braking decelerations of *0.4-0.6g* are typical, and greater than *1.0g* is possible. In what follows we consider vehicle acceleration $\pm a$ as a given value resulting from joint application of engine or braking forces, drag forces caused by wind resistance, rolling resistance, friction in the wheel bearings etc (Newcomb et al 1967; Wong 1993). We also assume that the most important internal degrees of freedom of the vehicle (corresponding to the pitch, bounce and wheel resonances are not excited by horizontal inertial forces produced during acceleration or deceleration. Being interested only in the low-frequency range of generated ground vibrations, we will model an accelerating or breaking vehicle of mass M as a point horizontal traction force $F_x = aM$ applied to the ground and moving along the road at vehicle speed. In fact, the resulting point force is the sum of partial forces applied to the ground from each wheel of the vehicle. For braking or accelerating respectively the direction of this force coincides with the direction of vehicle movement or is opposite to it. We also assume that adhesion between road and tyres is perfect. Thus, the limitations on acceleration (deceleration) caused by sliding or skidding will not be considered.

3.2 Governing equations and their formal solution

If a vehicle is accelerated or decelerated at constant absolute value a, then the point load force applied from a vehicle to the ground surface is a horizontal traction force that moves along the x-axis with a vehicle (see Fig. 5) and has the amplitude

$$F_x = aM . \tag{3.1}$$

The related mechanical shear stresses applied to the ground surface during acceleration or deceleration, i.e. during the period of time from $t = 0$ to $t = v/a$, where v is the final or initial speed, are described respectively as follows:

$$T_{xz}(\boldsymbol{\rho}, t) = -F_x\, \delta(x - at^2/2)\delta(y), \tag{3.2}$$

and

$$T_{xz}(\boldsymbol{\rho}, t) = F_x\, \delta(x - vt + at^2/2)\delta(y). \tag{3.3}$$

Here T_{xz}, are the components of a load stress tensor applied to the surface, $\boldsymbol{\rho} = \{x, y\}$ is the surface radius-vector, and $\delta(z)$ is Dirac's delta-function.

The ground vibration field generated by accelerating or braking vehicles in an elastic half space should satisfy the earlier written elastic Lame′ equation (2.5) and the boundary conditions on the ground surface which now take into account horizontal traction forces (3.2) and (3.3):

$$\begin{aligned} \sigma_{xz} &= 2\mu u_{xz} = -T_{xz}(\boldsymbol{\rho}, t), \\ \sigma_{yz} &= 2\mu u_{yz} = 0, \\ \sigma_{zz} &= \lambda u_{nn} + 2\mu u_{zz} = 0. \end{aligned} \tag{3.4}$$

The solution to the above problem can be written as the following Fourier integral:

$$u_i(\boldsymbol{r}, t) = (1/4\pi^2) \int_{-\infty}^{\infty}\int_{-\infty}^{\infty}\int_{-\infty}^{\infty} T_{xz}(\omega, \boldsymbol{k})\, G_{xi}(\omega, \boldsymbol{k}, z)\, e^{-i\omega t + i(\boldsymbol{k}\,\boldsymbol{\rho})}\, d\omega d\boldsymbol{k}\, , \tag{3.5}$$

where $\boldsymbol{r} = \{x, y, z\}$ is the three-dimensional radius-vector of the observation point, $\boldsymbol{\rho}(x,y) = \{x,y,0\}$ is the projection of the vector $\boldsymbol{r}$ onto the plane $z = 0$, $\boldsymbol{k} = \{k_x , k_y\}$ is the projection of a current three-dimensional wave-vector onto the plane $z = 0$, and ω is the circular frequency. Functions $T_{xz}(\omega, \boldsymbol{k})$ and $G_{xi}(\omega, \boldsymbol{k}, z)$ are respectively the Fourier transforms of the corresponding components of a load force source and the dynamic Green's tensor (Green's function) which are related to the space-time representations by the following formulae:

$$T_{xz}(\omega, \mathbf{k}) = (1/2\pi) \int_0^{v/a} \int_{-\infty}^{\infty} \int_{-\infty}^{\infty} T_{xz}(x, y, t)\, e^{i\omega t - i(\mathbf{k}\,\rho)}\, d\rho dt \qquad (3.6)$$

and

$$G_{xi}(\omega, \mathbf{k}, z) = (1/2\pi) \int_{-\infty}^{\infty} \int_{-\infty}^{\infty} \int_{-\infty}^{\infty} G_{xi}(\rho, t, z)\, e^{i\omega t - i(\mathbf{k}\,\rho)}\, d\rho dt\,. \qquad (3.7)$$

The Fourier transform of the relevant component $G_{xz}(\omega, \mathbf{k}, z)$ of the dynamic Green's tensor for an elastic half space is:

$$G_{xz}(\omega, \mathbf{k}, z) = i\frac{2k_x \nu_l \nu_t}{\mu F(k)} e^{-\nu_l z} + i\frac{k_x}{\mu F(k) k_t^2}[F(k) - 2k^2 b(k)] e^{-\nu_t z}\,. \qquad (3.8)$$

Here the following notation is used: $k = (k_x^2 + k_y^2)^{1/2}$, $b(k) = 2k^2 - k_t^2 - 2\nu_l\nu_t$, $\nu_{l,t} = (k^2 - k_{l,t}^2)^{1/2}$; $k_{l,t} = \omega/c_{l,t}$ are the wavenumbers of bulk longitudinal and shear acoustic waves; $c_l = [(\lambda+2\mu)/\rho_0]^{1/2}$ and $c_t = (\mu/\rho_0)^{1/2}$ are their phase velocities; $F(k) = (2k^2 - k_t^2)^2 - 4k^2\nu_l\nu_t$ is the Rayleigh determinant.

Using eqn (3.8) in eqn (3.5) and applying the equality $v_z(\mathbf{r},\omega) = -i\omega u_z(\mathbf{r},\omega)$, one can determine the spectrum of the vertical component of the ground vibration velocity which takes into account only Rayleigh waves:

$$v_z(\mathbf{r},\omega) = \left(\frac{2\pi}{k_R\rho}\right)^{1/2} \frac{\omega k_R^2 [k_t^2\nu_l\nu_t - k_R^2 b(k_R)]\cos\Theta}{\pi\mu F'(k_R) k_t^2} \times$$

$$T_{xz}(\omega, k_R\cos\Theta) e^{-k_R\gamma\rho} e^{ik_R\rho - i3\pi/4}\,, \qquad (3.9)$$

where $\Theta = \cos^{-1}(x/\rho)$ is the observation angle and $F'(k_R)$ is the derivative $dF(k)/dk$ taken at $k = k_R$.

Let us now discuss the Fourier transforms $T_{xz}(\omega, k_R\cos\theta)$ of the load forces (eqns (3.2) and (3.3)). Substituting eqns (3.2) and (3.3) into (3.6) and performing integration over x and y with taking into account properties of the delta-function, one can obtain the following expressions for accelerating and braking vehicles respectively:

$$T_{xz}(\omega, k_R\cos\theta) = -(1/2\pi) aM \int_0^{v/a} e^{i\omega t - ik_R\cos\Theta(at^2/2)} dt\,, \qquad (3.10)$$

$$T_{xz}(\omega, k_R\cos\theta) = (1/2\pi) aM \int_0^{v/a} e^{i(\omega - k_R v\cos\Theta)t + ik_R\cos\Theta(at^2/2)} dt\,. \qquad (3.11)$$

Taking these integrals analytically gives very bulky expressions containing functions of the complex variable $erf(z)$, which are inconvenient for numerical computations. The only useful conclusion derived from these expressions follows from their comparison. Such a comparison shows that they differ from each other only by the phase factor. Thus, if one is interested only in the amplitudes of generated vibrations, it is sufficient to consider just one of these cases, e.g. the case of braking vehicles. In what follows we will be interested mainly in 1/3 octave spectra of generated ground vibrations. This also allows us to use the approximate calculations of the integrals (3.10) and (3.11) without losing accuracy of calculations (for more details see Krylov 1996a,b).

3.3 Numerical calculations and discussion

The behaviour of 1/3-octave spectra of ground vibrations (in dB relative to the reference level of 10^{-9} m/s) generated by a braking lorry with $M = 20000\ kg$ for three different values of deceleration: $a = 1, 5$ and $9\ m/s^2$ is illustrated in Fig. 6. The initial speed is $v = 10\ m/s$ and $\Theta = \pi/3$. Other parameters are the following: mass density of the ground ρ_0 is $2000\ kg/m^3$, velocity of longitudinal bulk waves - $c_l = 471\ m/s$, shear waves - $c_t = 272\ m/s$ and Rayleigh surface waves - $250\ m/s$ (this corresponds to a Poisson ratio of $\sigma = 0.25$). The constant of ground attenuation of Rayleigh waves was set as $\gamma = 0.05$. It follows from Fig. 6 that amplitudes of generated ground vibrations at all frequencies increase with increase of a.

Similar calculations of the behaviour of generated ground vibration spectra for three different values of the initial speed (v = 5, 10 and 20 m/s) for a = 5 m/s^2 and $\Theta = \pi/3$ show that at frequencies higher than 4-5 Hz the corresponding spectra are almost independent of v.

The behaviour of generated ground vibrations (in *m/s*) as functions of acceleration a and initial vehicle speed v is illustrated respectively on Fig. 7 and Fig. 8. The ground vibration velocities were averaged over 1/3-octave band around central frequencies F = 4, 8, 16, 32 and 64 Hz. Note that oscillations of ground vibration level versus a take place for larger a, whereas oscillations versus v occur for smaller v. Also, the ground vibration velocity V is proportional to a for higher frequencies and for lower accelerations (Fig. 7) and does not depend on v for higher frequencies and higher v (Fig. 8). It is interesting that, according to Fig. 8, for low-frequency components of ground vibration spectra (4 and 8 Hz) the level of generated vibrations may differ by about 12 dB if initial speed v changes from 1 to 2 m/s. Thus, even small changes of v in this range may result in large difference in ground vibrations generated in infrasound frequency range. This may cause high statistical deviations in levels of ground vibration generated by slowly approaching vehicles.

4 CONCLUSIONS

Theoretical analysis of ground vibrations generated by vehicles travelling over traffic calming road humps and cushions shows that amplitudes and spectra of vibrations depend strongly on the shape of hump or cushion profile, relation between its height and length, vehicle speed and load, and ground elastic parameters. The developed theoretical model of generating ground vibrations agrees with the existing experiments and can be successfully used in practice.

Study of ground vibrations generated by accelerating and braking vehicles shows that for medium and upper bands of the spectra the generated ground vibration level is determined mainly by acceleration and is almost independent of the initial (final) vehicle speed. For low-frequency spectral bands, oscillations of ground vibration amplitudes versus both acceleration and initial speed may take place. These oscillations may be responsible for large statistical deviations of experimentally observed ground vibration levels.

5 REFERENCES

Ewing, W.M., Jardetzky, W.S. and Press, F. (1957). *Elastic waves in layered media.* McGraw-Hill, New York, N.Y.

Graff, K.F. (1975). *Wave motion in elastic solids.* Clarendon Press, Oxford.

Gutovski T.G., Dym C.L. (1976). Propagation of ground vibration: a review. *Journ. of Sound and Vibration*, **49**, 179-193.

Hanazato, T., Ugai, K., Mori, M. and Sakaguchi, R. (1991). Three-dimensional analysis of traffic-induced ground vibrations. *Journal of Geotechnical Engineering,* **117**, 1133-1151.

Hunt, H.E.M. (1991). Stochastic modelling of traffic-induced ground vibrations. *Journal of Sound and Vibration,* **144**, 53-70.

Krylov, V.V. (1995a). Comparison of two main mechanisms of generating ground vibrations by road vehicles. *Proc. I.O.A.*, **17**(4), 75-82.

Krylov, V.V. (1995b). Ground vibrations from accelerating and braking road vehicles. In: *Proc. 15th International Congress on Acoustics*, Trondheim, Norway, 26-30 June 1995, Ed. M. Newman, Vol. 2, p. 161-164.

Krylov, V.V. (1995c) Generation of low-frequency Rayleigh waves by heavy lorries. *Journ. Low Frequency Noise and Vibr.*, **14**(4), 165-172.

Krylov, V.V. (1996a). Computation of ground vibrations generated by accelerating and braking road vehicles. *Journ. of Vibration and Control (USA),* **2**, 299-321.

Krylov, V.V. (1996b). Generation of ground vibrations by accelerating and braking road vehicles. *Acustica-acta acustica*, **82**(4), 642-649.

Le Houedec, D. and Riou, Y. (1982). Analytic response of a semi-infinite soil mass to road traffic vibration, *Proceedings of the Conference on Soil Dynamics and Eartquake Engineering,* Southampton, pp. 209-233.

Newcomb, T.P. and Spurr, R.T. (1967). *Braking of road vehicles.* Chapman and Hall, London.

Taniguchi, E. and Sawada, K. (1979). Attenuation with distance of traffic-induced vibrations, *Japanese Society of Soil Mechanics and Foundation Engineering: Soils and Foundations* **19**, 15-28.

Watts, G.R. (1984). Vibration nuisance from road traffic - results of a 50 site survey. Transport Research Laboratory report LR1119 (Crowthorne).

Watts, G.R. (1998). Vehicle generated ground-borne vibration alongside speed control cushions and road humps. In: *Ground Dynamics and Man-made Processes: Prediction, Design, Measurement*, Ed. B.O. Skipp, Thomas Telford, London 1998, p. 87-101.

Wong, J.Y. (1993). *Theory of ground vehicles*, John Wiley & Sons, New York.

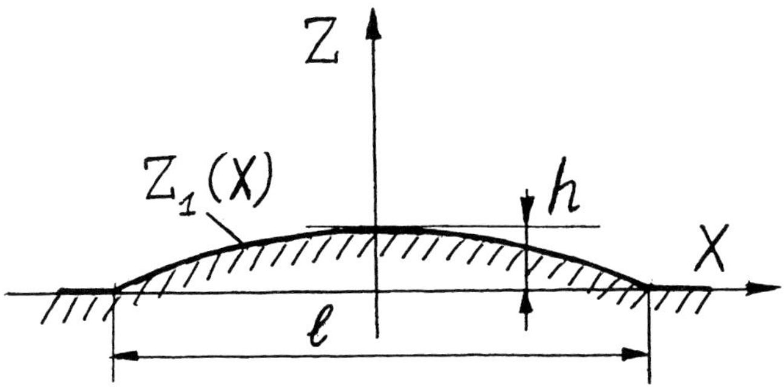

Fig. 1. Typical geometry of a road hump

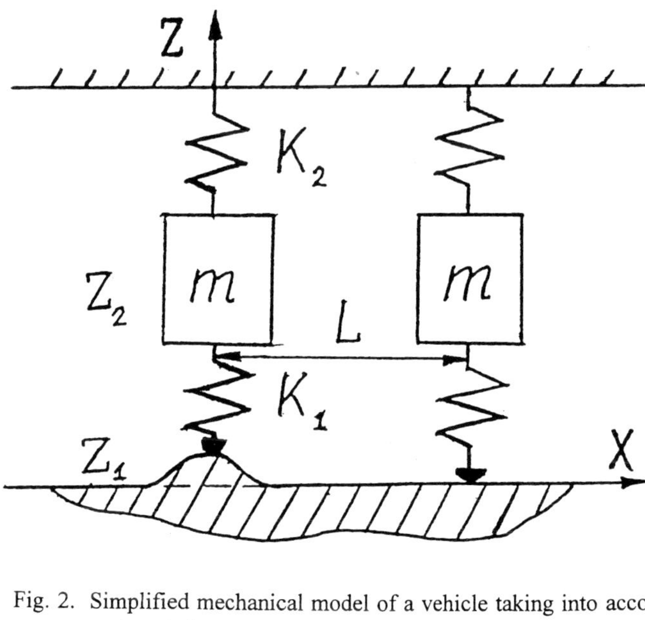

Fig. 2. Simplified mechanical model of a vehicle taking into account only axle hope resonances

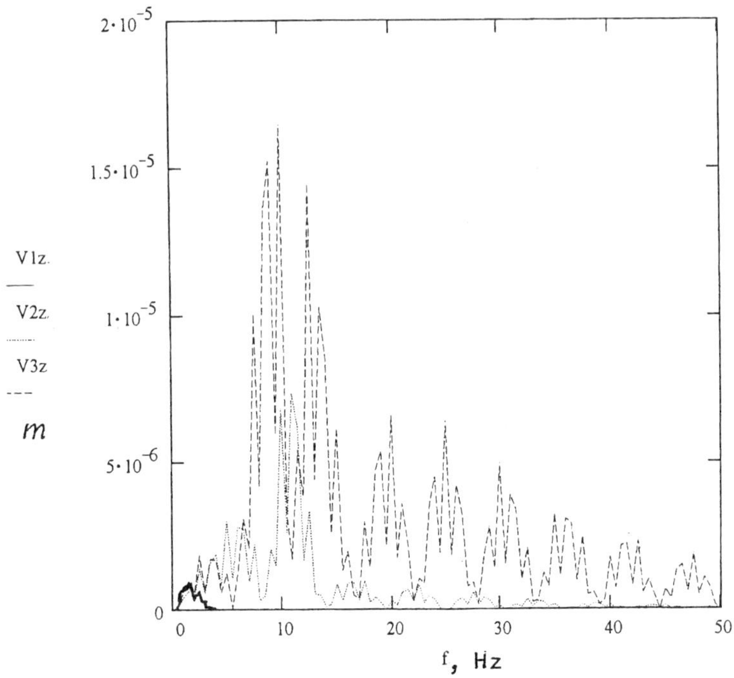

Fig. 3. Fourier spectra of ground vibrations generated by a vehicle travelling at speed $v = 5\ m/s$ over the three types of humps having the same characteristic length $l = 0.9\ m$ and height $h = 0.074\ m$: a Gaussian hump (curve V1z), a cosine-shaped hump (curve V2z) and a rectangular hump (curve V3z)

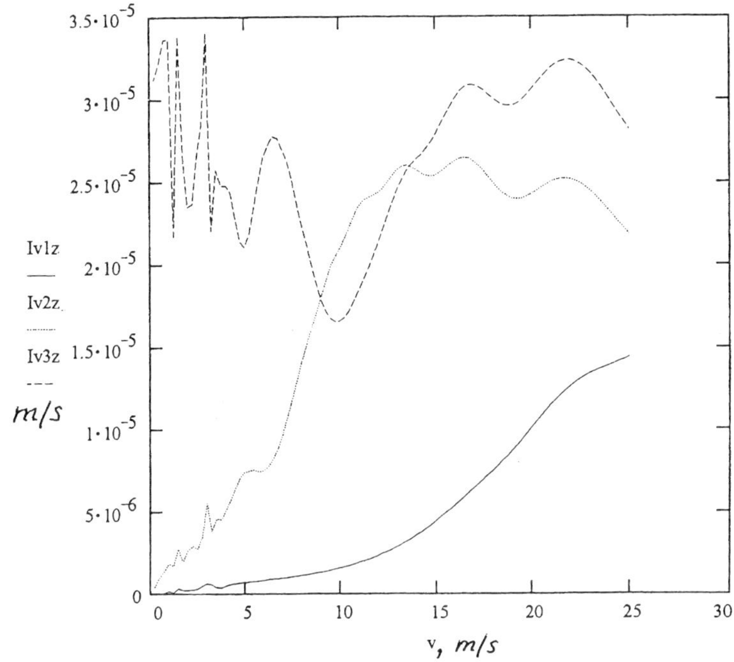

Fig. 4. Integral values of vehicle-generated ground vibrations as functions of vehicle speed for the three types of humps having the same characteristic length and height: a Gaussian hump (curve Iv1z), a cosine-shaped hump (curve Iv2z) and a rectangular hump (curve Iv3z). Parameters of the humps are the same as in Fig. 3

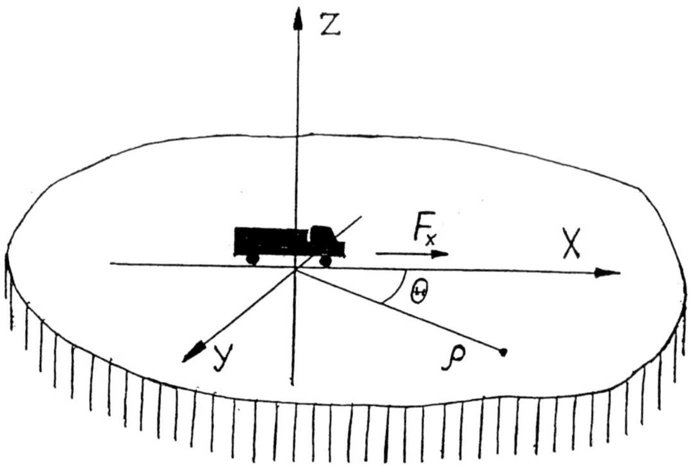

Fig. 5 Traction force F_x applied to the ground during vehicle braking

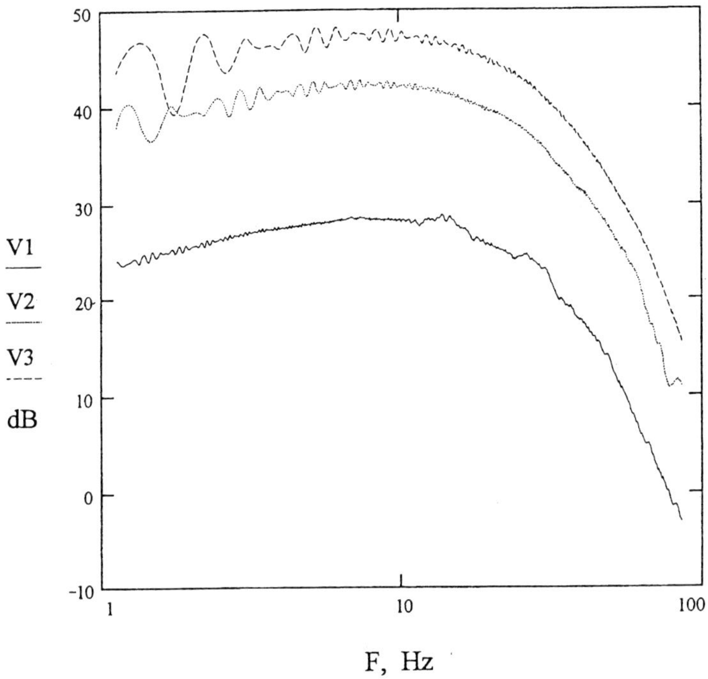

Fig. 6. 1/3-octave ground vibration spectra calculated for three values of vehicle acceleration (deceleration) *a*: $a = 1\ m/s^2$ (curve *V1*), $a = 5\ m/s^2$ (curve *V2*) and $a = 9\ m/s^2$ (curve *V3*)

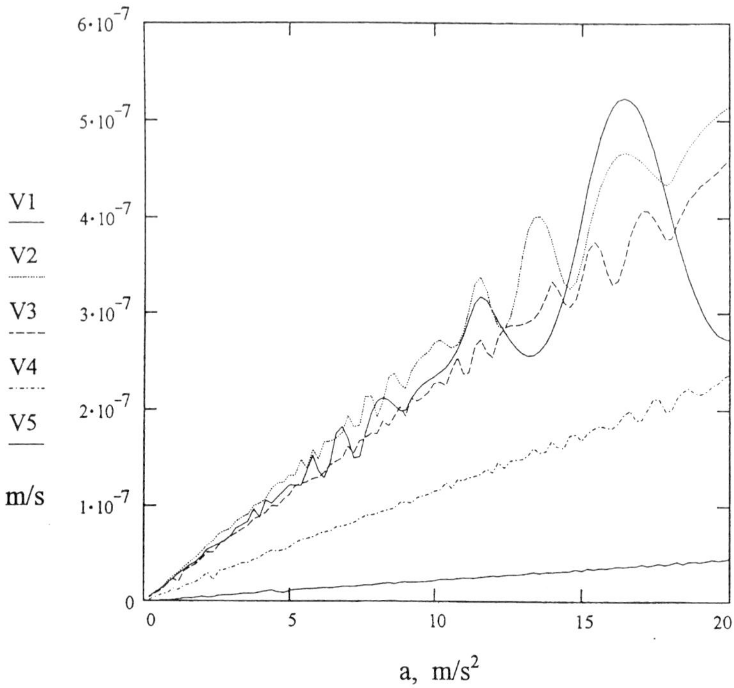

Fig. 7. Ground vibration amplitudes (in *m/s*) averaged over 1/3-octave frequency band as functions of vehicle acceleration (deceleration) *a* for the central frequencies *F = 4,8,16, 32 and 64 Hz* - curves *V1, V2, V3, V4* and *V5* respectively

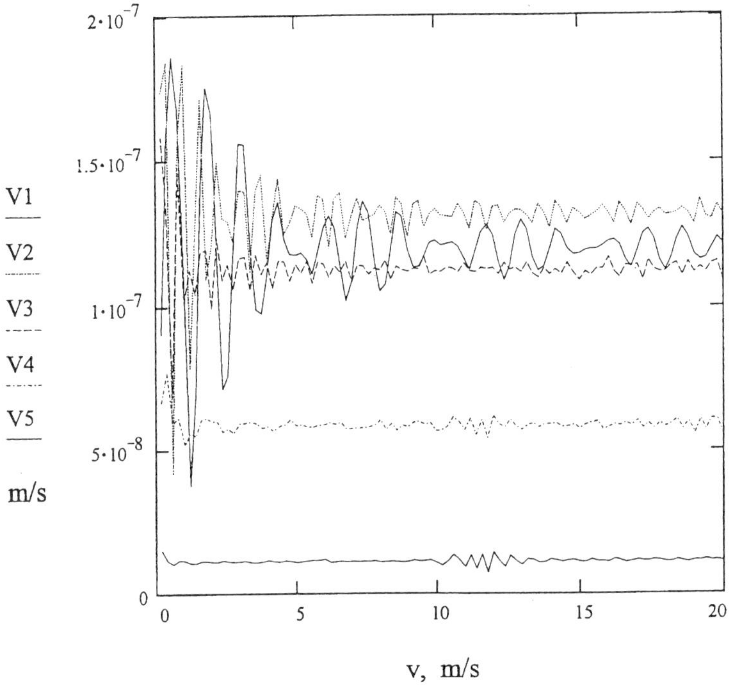

Fig. 8. Ground vibration amplitudes (in *m/s*) averaged over 1/3-octave frequency band as functions of final (initial) vehicle speed *v* for the central frequencies *F = 4,8,16, 32* and *64* Hz - curves *V1*, *V2, V3, V4* and *V5* respectively

S682/006/99

European co-ordinated railway noise research

B HEMSWORTH
European Rail Research Institute, Utrecht, The Netherlands

1 INTRODUCTION

Current EU policy is to support movement of passenger and freight traffic by rail. It is also policy that environmental noise levels should not increase. The dilemma facing the railways is to provide environmentally acceptable railways so that these apparently conflicting policies can be implemented without increasing the cost.

In principle, noise can be reduced by rolling stock improvements and track improvements. The trade off between noise level from an individual train and L_{eq} is such that if the noise from every train could be reduced by, say 10 dB(A) a ten fold increase in traffic could be tolerated before the original L_{eq} was exceeded. In reality it is likely that in the short to medium term major noise reductions at source are only feasible for freight vehicles. The overall potential effect on environmental noise levels will therefore depend on the balance between passenger and freight vehicles. Despite this the rewards for running quieter trains are potentially high for the environment and the rail operators.

Lineside noise barriers can be effective and are integral parts of the noise control strategy for most new railway schemes. However, they present a potential hazard to track workers, adversely affect mechanised track maintenance and are a permanent visual intrusion to railway passengers and to lineside residents. The visual intrusion will become more serious if the current trend in Europe continues for building noise barriers in excess of 4m height. Cost benefit studies have highlighted the poor return from using barriers, particularly high ones.

Additionally, national legislation, particularly for new schemes, often requires secondary sound insulation to be provided for properties where specified trigger noise levels are exceeded. This does not reduce the external noise level but provides protection when inside the home. In the UK, entitlement to sound insulation for new roads and railway infrastructure is dealt with under the Land Compensation Act 1975 and the associated Statutory Instruments.

Recent research has identified potential quieter designs that could be applied to vehicles and track. It seems possible, therefore, that there is a possibility of achieving the policies with a combination of these measures but each will have an associated cost. To determine the most appropriate strategy requires a balance to be made between the needs of the environment, the limitation in the effectiveness of any solution and the cost of their introduction. It would be counter to the "transfer to rail policy" if the costs associated with the use of low noise technology are such that the railways' commercial competitiveness is compromised.

2 NOISE MITIGATION AT SOURCE

At current speeds in Europe the prime noise source is the wheel rolling on the track, as speeds increase from today's levels, aerodynamic noise will become more significant and this noise source currently attracts research funding in Japan. The dominance of aerodynamic noise in Japan is also a consequence of extensive use of trackside noise barriers which are effective in reducing noise sources at rail height (rolling noise) but will not affect higher sources (aerodynamic noise).

For the last 30 years or so European member railways of the International Union of Railways (UIC) have carried out co-operative and locally funded research on rolling contact noise. Through these studies a greater understanding of the noise processes has been obtained and this understanding has enabled a number of noise mitigation solutions to be developed. The background to this work has been the development of the TWINS model for the generation of rolling noise and the flow diagram is shown in Figure 1.

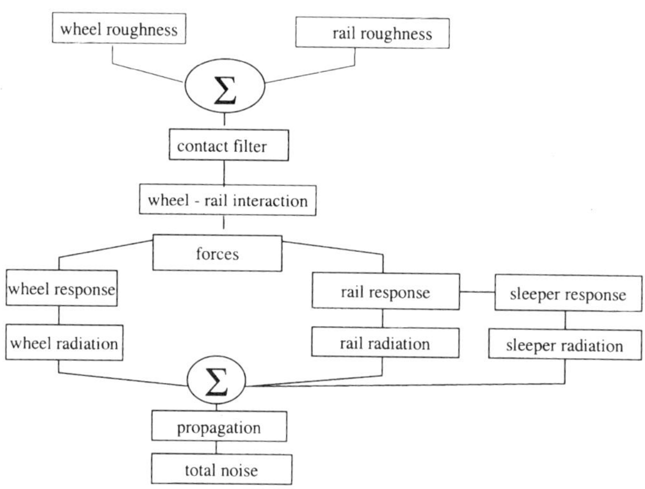

Figure 1 Flow Diagram for TWINS rolling noise model

Rolling noise generation is a forced vibration process with the force generated in the rolling contact zone causing wheels, rails and sleepers to vibrate and hence generate noise. The model assumes that, in the frequency regime, there is a linear relationship between generated noise and the combined roughness of wheel and rail. Smooth surfaces are therefore a necessity for a quiet railway. Wheel roughness is controlled by the braking system and the use of cast iron tread brakes has, in the past, been incompatible with smooth wheels. For uncorrugated track, wheel roughness dominates rail roughness, this results in cast iron tread braked vehicles being some 10 dB(A) noisier than disc braked vehicles at the same speed.

Cast iron tread braked freight vehicles are still the norm for the majority of Europe and there are currently no approved alternative braking systems that can be used for International freight traffic.

In broad terms the contributions to total noise from wheel and track are approximately equal. This balance changes with detail design of wheel and track but the important consequence is that significant noise reduction can only be achieved if the contribution from both sources is reduced.

Low frequency noise is generated by the sleeper, high frequency noise is generated by the resonant response of the wheel, while the rail dominates the mid frequencies and contributes at all other frequencies.

Assessment of the noise generation and radiation processes of Figure 1 indicates that rolling noise can be reduced by the following actions.

- Reduction of force (smooth wheels and rails)
- Reduction of vibration (damping, shape)
- Reduction of radiation efficiency (size, shape)
- Increase of propagation losses (shrouds and barriers).

These actions form the background to a number of current noise investigations.

3 SUMMARY OF PROJECTS

A number of projects were proposed in 1995 as part of the EU Brite Euram Programme of the EU Research Directorate (DG XII) and were aimed at reducing railway noise at source. In Europe freight traffic is perceived as the main noise problem and the proposals focussed on noise reduction from cast iron tread braked freight vehicles operating on ballasted track. Consortia were formed with key experts from railways, industry, research institutes and universities to launch the following projects.

- Silent Freight (low noise solutions for freight vehicles)
- Silent Track (low noise solutions for railway track) and
- Eurosabot (methodologies for the design of low noise tread brakes).

Silent Freight commenced in 1996 followed by Silent track in 1997. The objectives were to produce economic solutions, which will reduce noise from wheel/rail interaction, wheels, vehicle superstructure and track. The combined aim, was to reduce rolling noise from freight vehicles by up to 10 dB(A) by developing innovative solutions that can be applied through retrofit to existing vehicles and infrastructure or to future designs.

The general approach has been similar for both projects. Modelling has quantified the dynamic properties of components necessary to achieve the required reduction, followed by laboratory/rig tests to confirm that the required values had been achieved and that components responded in the predicted manner. This was then to be followed by full scale testing of prototype designs. The final testing stage took place at the Czech Republic Test Centre at Velim in June/July 1999. Preliminary analysis of the results indicates are that in the majority of cases the expected results have been achieved and they are presented below.

Eurosabot commenced in 1996 with the objective of increasing the understanding of the generation and growth processes of wheel surface roughness during braking and thus developing a methodology for designing tread brake solutions which would lead to lower levels of wheel roughness and noise. This could be considered as a fundamental research project that would develop design rules.

In 1997 UIC initiated a plan to reduce freight vehicle noise levels by replacing cast iron tread brake blocks with blocks made from composite materials. From past experience these would be expected to give rise to lower roughness levels on the wheels and hence reduce noise. Low stress wheels would need to be incorporated into the design to combat the increase in residual stress levels in the wheel caused by the use of composite brake blocks.

The following sections describe the techniques that have been used in these projects and summarise the potential consequential noise reduction. It must be stressed that a number of engineering issues involving safety, maintenance and inspection need to be resolved before the prototype designs can be taken forward to the next stage of implementation.

4 POTENTIAL NOISE REDUCTION

4.1 Smooth wheels and rails

Since the force input is proportional to the sum of wheel and rail roughness, if wheel roughness is to be controlled by a change of braking design, rails also need to be retained in as smooth a condition as possible. The noise effects of corrugated (high roughness) rails are well known.

As part of the Silent Track project a number of models for rail roughness growth are being developed. These are being used to confirm that there will be no adverse roughness effects caused by the use of low noise track components.

In Eurosabot success has been achieved in that it has been shown that thermal instabilities, material transfer and block vibration are important factors in the generation of surface roughness. The rig tests indicated a number of promising solutions but benefits obtained from those tests have not been achieved on the full scale field tests carried out to date.

UIC tests in 1999 on the Gotthard Pass indicated that noise reductions of between 8 and 10 dB(A) could be achieved from the particular composite brake blocks tested, but further work is necessary to confirm a braking performance, under all conditions, that would be acceptable before approval for their use in international traffic can be obtained.

This range can be used as a working estimate for the potential noise reduction that can be achieved from a change of braking system.

4.2 Low noise wheels

The wheel can be considered as a circular plate with radial and axial modes of vibration. The resonant response needs to be reduced. The first stage of the design concerns shape optimisation where the radiating radial and axial modes are decoupled. This requires the cross section to be as symmetrical as possible which implies straight webs or slightly curved webs. All wheels developed for Silent Freight have been shape optimised in so far as the appropriate UIC leaflets allow for residual stress considerations.

The resonant response can also be reduced by adding damping. Within Silent Freight, damping has been added to the wheels by using tuned absorbers.

Further options tested have been, perforated wheels (low frequency reduction) in combination with ring dampers (high frequency reduction) and wheels with screened webs.

4.3 Low noise track components

The rail behaves as a vibrating beam supported at discrete intervals by sleepers. Noise mitigation will be achieved by reducing the length of vibrating rail. A controlling influence on track vibration decay is the stiffness of the rail pad and modelling studies have indicated that pads with high dynamic stiffness (800 kN/mm) are near optimum for minimising the track component of noise. This is at variance, however, with the requirement for lower stiffness pads to protect sleepers and for rail roughness growth considerations.

Damping can also be added and prototype track designs for Silent Track included high stiffness rail pads in combination with tuned absorbers. This option is considered to be a potential retrofit solution. The effect of stiffer pads and damping was also assessed separately.

It has also been shown that vibration of the rail foot in bending generates significant noise. A modified rail cross section has been produced where the rail foot of UIC 60 rail has been reduced from 150mm to 100mm. To maintain stability a new rail fastener has been produced where the rail is supported under the head.

In combination with tuned absorbers and modified sleepers this design has been designated "new track".

Table 1 shows the expected reduction of rolling noise from different combinations of low noise wheel and track components developed in the projects.

	WHEEL					
TRACK	**Ring damper**	**Perforated wheel**	**Perforated wheel + ring**	**Shape optimised wheel**	**Shape optimised + absorber**	**shape optimise + web shields**
Ref track + stiff pads	3	3	3.5	3	3.5	3.5
Ref. Track + tuned absorber	5	6	6	6	6	6
Ref. Track + Pads + tuned absorber	5.5	6.5	7	6.5	7	7
New track	2	2.5	2.5	2.5	2.5	2.5
new track + tuned absorber	6.5	7.5	7.5	7.5	8	8

Table 1 noise reduction in dB(A) for various wheel/track combinations vs. reference

The data in the table indicate that a reduction in total rolling noise of between 6 and 8 dB(A) can be achieved when damped wheels and rails are part of the mitigation.

4.4 Bogie shrouds and barriers

Following encouraging results from national projects in the UK, Italy and Germany a bogie shroud and low trackside barrier combination was also designed and tested as part of Silent Freight and Silent Track projects.

To achieve a maximum reduction, it is necessary for the bottom edge of the bogie shroud to be as close as possible to the head of the rail and to overlap the top of the barrier. This must be achieved whilst complying with gauging restrictions.

The gauging requirements for a freight vehicle which could be used in international traffic do not allow this and the combination tested within these projects gave a noise reduction of about 3 dB(A).

4.5 Combination of measures

In summary, modifications of tread braking resulting in smoother wheels are expected to produce reductions of between 8 and 10 dB(A

Low noise wheel and track components are expected the reduce the noise by 6 – 8 dB(A)

Bogie shrouds and low barriers provide a reduction of about 3 dB(A)

As a first approximation it can be assumed that for combinations of the above options, these results can be added together which suggests that if only acoustic considerations are taken into account noise reductions of up to about 20 dB(A) are achievable for cast iron tread braked freight vehicles. The braking system to provide smooth surfaces may not allow all the low noise designs tested in these projects to be implemented. That figure may therefore not be achievable in practice.

It is possible to choose a mix of options to achieve intermediate targets, and the order in which mitigation is applied is at the discretion of the user, although logic would suggest that the most appropriate is as presented above. This indicates smooth wheels as a first modification with additional benefit, if required, from the use of low noise wheel and track designs and finally bogie shrouds if additional reduction is absolutely necessary.

5 COST BENEFIT STUDIES

As described above a number of alternative low noise wheel and track designs are potentially available to give noise reductions up to 8 dB(A). These designs are currently only in the pre prototype stage and much work is needed before they can be turned into railway engineering hardware and indeed, it has yet to be shown whether all the designs are capable of implementation, when safety, maintenance and inspection issues are included.

An economic study is to be carried out which will assess the costs of costs and benefits balance of different options in comparison with the use of noise barriers and sound insulation of lineside property.

A similar study has just been completed for UIC in which a prediction was made of the noise benefits arising from the introduction of different low noise freight vehicles on two major international freight routes which span Holland, Germany, Switzerland, Italy and France. Although considered to be significant railway freight routes, mainline and local passenger services also operate on the same routes and their noise contribution was included in the assessment. Additionally, where specific noise targets were exceeded, noise barriers or track mitigation options were introduced. Finally, sound insulation was applied to those properties where noise levels exceeded the targets even after active and passive mitigation was included. Costs for each of the measures were also determined.

The studied scenarios included freight rolling stock modifications giving noise reductions relative to cast iron tread braked freight vehicles of 5 dB and 10 dB and track modifications causing reduction in noise from all vehicles of 5 dB. No modification was made to passenger vehicles.

When factored to provide a preliminary assessment on a European scale the major conclusions can be summarised as follows:

- Solutions containing rolling stock improvement prove to be optimal. They are cost effective alone and in combination with other measures.
- Scenarios containing 4m noise barriers did not have an additional benefit to compared with 2m solutions.
- Track improvements are only interesting in combination with rolling stock improvement. This conclusion could change if higher benefits from track modification become available.
- Additional cost reduction may be achieved by a further improvement in rolling stock.

- If only indoor noise protection is used as a criterion, insulated windows are less expensive than noise barriers.

Thus the introduction of low noise vehicles was shown to have environmental as well as cost benefits. The economic study of Eurosabot, Silent Track and Silent Freight will assess specifically the noise mitigation developed within those projects and will provide more detailed background of the costs and benefits associated with the alternative vehicle and track options.

6 CONCLUSIONS

A number of low noise design options for rail vehicles and track have been developed and are now at the prototype stage. The next stage will be to consider the safety, maintenance, inspection and cost/benefit aspects of the designs leading to possible implementation.

7 ACKNOWLEDGEMENTS

The author would like to thank ERRI for permission to publish this paper and UIC for their continued support of railway noise research. Thanks are also due to the Silent Freight, Silent Track and Eurosabot partners for their hard work, perseverance and commitment to the projects.

S682/007/99

A review of computer modelling of sound in buildings

B M SHIELD, S DANCE, and **L N YANG**
School of Engineering Systems and Design, Southbank University, London, UK

1 INTRODUCTION

Over the past twenty years there has been increasing use of computer models for the prediction of noise in enclosed spaces. Originally the predominant use of models was for the prediction of industrial noise levels, to enable acoustic consultants, architects and noise control engineers to determine whether or not a proposed building would comply with the noise at work guidelines and legislation. The original models were capable only of predicting noise levels in diffuse spaces, that is spaces with an even distribution of sound energy, such as is found in spaces of regular proportions with an even distribution of absorptive material. However, increasing computer power, memory and speed has facilitated increased sophistication of models, so that it is now possible to predict a wide range of acoustic parameters in addition to noise levels. Thus, in addition to being used for simulation of industrial spaces, many computer models are now available for use by architects and acoustic consultants in the design of buildings such as theatres and concert halls.

This paper presents a review of computer modelling techniques for the prediction of the acoustics of enclosed spaces. Models that have been developed at South Bank University in the last ten years are described, together with the types of space for which they are applicable and their expected accuracy.

2 TYPES OF MODEL

The original models tended to be empirical models in which the calculation of noise levels were based upon relatively simple equations derived from classical acoustics theory [1-3]. However, the application of such models was limited to diffuse spaces and the prediction of noise levels, in linear dB or in octave or third octave bands only. As computers became more sophisticated it was possible to incorporate more complex mathematics into the models and over the last 10 years there has been a profusion of models developed, both for research purposes and as commercial products.

There are currently two broad categories of computer model for the prediction of room acoustics: ray tracing and image source models. These models are based upon geometric acoustics, in which the calculations involve summation of sound energy, rather than wave theory.

In ray tracing models a source is assumed to emit a predetermined number (up to several thousand) of rays, depending upon the volume of the space and the acoustic parameters to be predicted. Energy is assigned to each ray depending upon the sound power output of the source. Each ray from each source is traced around the space, losing energy through absorption and scattering as it encounters room surfaces and objects in the room (fittings). The passage of a ray is terminated when the energy it carries has reduced to a negligible level. The receiver points at which the sound is to be determined are represented as small volumes within the space. Each receiver accumulates the energy currently carried by each ray passing through it. The ray tracing technique has been further developed to the beam tracing method, in which a cone or pyramid shaped volume of rays is traced, rather than individual rays.

In image source models each reflection off a room surface is treated as if it was a ray emitted from the reflection of the source in that surface. Multiple reflections off more than one room surface are assumed to be emitted by an image source which has been reflected in many surfaces. Image source models perform well in regular shaped spaces but the construction of image sources makes them very time consuming in irregular shaped spaces.

A limitation of both ray tracing and image source models is that as they are based upon geometric acoustics and energy only is simulated, it is not possible to predict phenomena such as standing waves and interference effects. The only way to achieve this is to use models based upon wave theory, taking into account phase and pressure amplitude of the sound waves. However, such models are very time consuming in terms of computer running time and as yet are not practicable for every day use.

3 ACOUSTIC PARAMETERS PREDICTED

As stated above, early models were used to predict sound levels only in a space, the calculations of these being based upon energy summation. In a diffuse space it was also possible to predict reverberation time, based upon the classical acoustic formulae of Sabine and Eyring.

With current models it is possible to predict the impulse response at any receiver point in both diffuse and non-diffuse spaces. The impulse response shows the intensity and time of arrival at the point of the direct sound from a source, and successive reflections off room surfaces. Many acoustical parameters which are important for sound quality and speech intelligibility, including reverberation time, can be calculated from the impulse response. In addition to reverberation time, parameters commonly predicted include early decay time, EDT, which is the time taken for the first 10 dB decay of the sound; the clarity index, C80 (or C50), which is the ratio of the sound energy arriving in the first 80 ms (or 50 ms) to the sound energy received after that time; and deutlichkeit, or definition, which is the ratio of the sound arriving within the first 50 ms to the total sound energy.

Another parameter used specifically to describe speech intelligibility in a space is the speech transmission index, STI. The value of this parameter lies between 0 and 1, the closer it is to 1 the better the speech intelligibility in the space.

4 VALIDATION OF MODELS

It is vitally important, when models are developed or marketed, that a wide ranging programme of validation is undertaken. Models should be tested in as many different types of space as possible. In the case of models for industrial spaces, they should be tested in diffuse and non diffuse spaces, both empty and fitted, with varying numbers and configurations of fittings. Models should also be tested in spaces with little absorption as well as spaces with highly absorbent surfaces. All of the relevant acoustic parameters should be predicted, to ensure accuracy over as wide a range of parameters as possible.

It is usual to assume that an 'accurate' model for sound level prediction is one in which sound levels are predicted to within 2 dB of actual levels. For temporal parameters such as reverberation time, accuracy of within 10% – 14% is usually taken to be acceptable.

5 MODELS DEVELOPED AT SOUTH BANK UNIVERSITY

Over the last 10 years research has been carried out at South Bank University into the development of various types of computer model for the prediction of the sound field in many kinds of enclosed space. Ray tracing, image source and pressure based models have been developed for the prediction of sound levels and acoustic parameters in diffuse and non-diffuse spaces. The model developments have included extensive validation in a wide variety of spaces. Spaces for which models have been developed and tested include factories, offices, concert halls, recording studios, atria, laboratory spaces, scale models and underground stations. The industrial spaces modelled have included fitted and non-fitted spaces, and rooms containing a barrier.

In the course of the research all of the required acoustic parameters have been predicted accurately in many spaces. For industrial spaces it is usual to predict sound levels and reverberation time only. In spaces that are used for music and speech it is important to know the value of temporal parameters such as EDT and clarity indices. The main aim of the work on long enclosures was to predict parameters relating to speech intelligibility, including the speech transmission index.

Current work involves the simultaneous prediction of internal and external sound levels, together with the development of a database of absorption data which has been shown to work in all types of computer model.

6 OPTIMISATION OF MODELS

In addition to development and validation aimed at improving the accuracy of models, considerable investigations have been carried out into optimisation of the models [4,5]. The aim of this was to make them as efficient as possible, at the same time increasing their 'user-friendliness', by reducing the amount of input data required, and the running time, while maintaining prediction accuracy. Among the factors that have been investigated are the ways in which the geometry of a space is represented; the representation of fittings; reduction of number of rays required; and receiver size. The input data that is critical to accurate predictions has been determined. It has been found that it is possible to accurately predict the sound field in typical fitted industrial spaces while reducing the running time from hours to minutes, and halving the amount of input data required.

7 MODELS FOR THE PREDICTION OF INDUSTRIAL NOISE

A ray tracing model and an image source model were developed for the prediction of industrial noise. These models were designed to predict sound levels around a factory at different frequencies. The models were tested using both laboratory data and data measured in a range of factories of different shapes and sizes.

The current ray tracing model, FAME [6] is a further development of models which were originally based on the Ondet and Barbry ray tracing model [7] for the prediction of sound levels in empty and fitted factory spaces [8]. In addition to a ray tracing model an image source model, CISM, was also developed [9].

Both models were originally designed for the prediction of sound levels in industrial spaces only. The models were tested using data collected in a laboratory space with a single source in nine different configurations of fittings. Thirteen industrial halls, of volumes ranging from approximately 5000 m^3 to 40000 m^3, and with from one to twenty-one machines, were also used in the validations. In all cases both models were found to predict sound pressure levels with an average error of 2 dB or less [8-10].

7.1 Modelling of barriers

In addition to sound level predictions both models have been used to predict insertion loss around a barrier. Various different methods of modelling a barrier were tested and the predictions resulting from each one compared with measurements made in a factory and in a test room. The method found to give the most accurate results involved modelling the barrier as an infinitely thin absorptive plane, surrounded by an additional imaginary plane one or two wavelengths in depth, to form a diffraction area. If a ray strikes the diffraction area it is randomly redirected with no additional attenuation. This method was tested in six arrangements of the two spaces containing a barrier [11,12]. The ray tracing model FAME was accurate to within 2 dB of measured insertion loss at all frequencies. However, it was not possible to achieve accurate results with the image source model CISM, except at high frequencies, as the complex geometry could not be realistically represented by the image source model.

7.2 Modelling of fittings

Fittings may be represented individually, or grouped into zones. The most efficient representation, in terms of accuracy, run time and input requirements, has been determined. It was found that, for industrial halls, the simplest and most effective representation of fittings is as a zone covering the floor, a more detailed description only marginally improving accuracy at the expense of considerably increased user input time [13].

8 PRESSURE BASED MODEL

In order to predict the effects of interference on sound levels, which can be significant in certain circumstances, a model was developed which is based upon pressure and takes into account phase differences between direct and reflected sound. The model, PHASE, has been validated using data measured in three different situations where interference effects were measured: a laboratory space where the sound level was found to vary significantly with receiver height; an empty factory space where interference effects were thought to be due to the reflections from the hard floor; and small rooms representing an empty factory. The original model correctly predicted the trend of sound propagation, showing that interference effects occurred at low frequencies, although precise agreement between predicted and measured values was not obtained [14]. However, with the increase in computer power in recent years, the model has been improved such that it now accurately predicts the sound level, including interference effects, in all octave bands [15].

9 MODELS FOR THE PREDICTION OF THE SOUND FIELD IN PERFORMANCE SPACES

The main difference between the acoustic characteristics of performance spaces and those of industrial spaces is that the former tend to have surfaces which are more acoustically absorbent than the latter.

Both FAME and CISM were extended to predict the impulse response at receiver points. From this it is possible to calculate parameters such as RT, EDT, and C80, which are important for sound quality in spaces where speech or music are performed. The models have been tested in various different types of space including factories, offices, a concert hall, and a recording studio. To achieve accurate prediction of the sound field in the performance spaces it was found necessary to use absorption data based upon the Millington reverberation time formula, rather than the classical formula of Sabine. With this data it was possible to accurately predict sound pressure levels and reverberation time in all the spaces [16].

10 SIMULTANEOUS INTERNAL AND EXTERNAL PREDICTION

Previous computer models have predicted the internal and eternal sound fields separately. FAME has recently been extended to predict the internal sound field, while simultaneously predicting the sound levels around a building due to an aperture in the building façade. The method for the calculation of external noise is based on the diffraction/barrier model, each surface of the building being extended by a transparent diffraction area. Data previously measured in an anechoic chamber, using a box with an open aperture of varying size to simulate a building in a free field, was used to validate the model. FAME was found to give accurate predictions of sound pressure levels around the box. The next stage in the model development is to validate the model using data measured in and around real buildings.

11 DEVELOPMENT OF ABSORPTION DATABASE

In the course of the research described here, it has become obvious that there is a need for a standard set of absorption data on common construction materials, for use with all types of model. A database of absorption coefficients has therefore been developed, consisting of those coefficients which have been found to give accurate predictions of a range of acoustic parameters using different types of model, including commercial models, in a variety of buildings and spaces [17].

12 PREDICTION OF THE SOUND FIELD IN LONG ENCLOSURES

It is known that long enclosures such as underground stations are non-diffuse spaces [18]. With a single source at one end of a long enclosure, the sound level decreases with increasing distance from the source, while the early decay time increases. Two ray tracing models have been developed in order to predict the sound field in underground stations, with particular attention being given to the prediction of speech intelligibility [19-21]. One model is designed to predict the acoustics of underground stations of rectangular cross section, and the other of stations with circular or semi –circular cross section. Sound pressure levels, early decay time, clarity and definition are predicted throughout the space, together with the speech transmission index.

This is the first time that ray tracing has been used to model curved surfaces exactly, previous models having approximated a curve by a series of planes. However, in the course of development of the model it was found that, even with a large number of planes, it is not possible to accurately predict focussing effects that may occur due to curved surfaces [22].

The models have been validated using data measured in a real station of rectangular cross section, and in a scale model of a station of circular cross section. In the scale model only sound pressure levels and early decay times along the length of the station were measured and predicted in octave bands from 125 Hz to 1000 Hz (full scale). Good agreement was obtained at all frequencies, apart from the 125 Hz band. In the rectangular station the parameters EDT, sound pressure level, C50, and D50 were measured and predicted in octave bands from 125 to 4000 Hz. The speech transmission index was also predicted and predictions compared with measured values. Good agreement was found between predictions and measurements at most frequencies with the sound pressure level predictions being within 1.5 dB of measured levels on average, and the average errors of EDT, C50 and D50 being less than 13%, 2% and 8% respectively. The predicted overall STI values were very accurate, with errors of around 0.05 at all frequencies.

13 CONCLUSIONS

In the past ten years there have been many developments in the field of computer modelling of the acoustics of buildings, many of which have become possible owing to the increase in the capabilities and speed of computers. This has enabled the development of many different kinds of models, of increasing sophistication. Today's models can accurately predict a wide range of acoustic parameters in many different types of buildings.

Researchers at South Bank University have contributed significantly to these developments, by writing new models, improving their accuracy and efficiency, and continually testing models in many different scenarios.

The most important aspect of a computer model for predicting the sound field is its accuracy. Ease of use is another important factor - a highly accurate model that is difficult to use, or takes many hours to run, is of little practical value. Similarly, a model with highly developed graphics for input and output is worthless if the results produced do not represent the real world. It is therefore important for users of models to be aware of how they have been developed and validated, in order to be assured of their accuracy, and their practical application.

14 ACKNOWLEDGEMENTS

Much of the research described here has been funded by the Engineering and Physical Sciences Research Council.

15 REFERENCES

[1] R Friberg. Noise reduction in industrial halls obtained by acoustical treatment of ceilings and walls. Noise control and Vibration reduction, March 1975.
[2] B M Shield. A computer model for the prediction of factory noise. Applied Acoustics 13, 471-486, 1980.
[3] C Jones and B M Shield. A computer model for the prediction of noise levels in factories. Proc. Institute of Acoustics 10(5), 21-30, 1988.
[4] S M Dance and B M Shield. The effect on prediction accuracy of reducing the number of rays in a ray tracing model. Proc Inter-noise '94, 2127-2130, 1994.
[5] S M Dance and B M Shield. The prediction of reverberation time by ray tracing in non-performance spaces using a reduced time resolution. Proc. Inter-noise 97, 779-782, 1997.
[6] S M Dance, J N Pinder and B M Shield. The prediction of reverberation time in fitted rooms. Proc. Euronoise 98, 667-670, 1998.
[7] A M Ondet and J L Barbry. Modelling of sound propagation in fitted workshops using ray tracing. J. Acoustical Society of America, 86(1), 172-178, 1989.
[8] S M Dance. The development of computer models for the prediction of sound distribution in fitted non-diffuse spaces. PhdD thesis, South Bank University, 1993.
[9] S M Dance and B M Shield. The complete image-source method for the prediction of sound distribution in non-diffuse enclosed spaces. Journal of Sound and Vibration 201(4), 473-489, 1997.

[10] B M Shield and S M Dance. The optimisation of a ray tracing model for the prediction of sound fields in enclosed spaces. Submitted to J. Sound and Vibration, 1999.
[11] S M Dance, J P Roberts and B M Shield. Computer prediction of insertion loss due to a single barrier in a non-diffuse empty enclosed space. J. Building Acoustics 1(2), 125-136, 1995.
[12] S M Dance and B M Shield. A refined diffraction approximation for free standing barriers in enclosed spaces. Proc. Internoise 98, paper 126, 1998
[13] S M Dance. The representation of fittings in the computer prediction of sound propagation in enclosed spaces. Applied Acoustics 52(1), 71-83.
[14] S M Dance, J P Roberts and B M Shield. Computer prediction of sound distribution in enclosed spaces using an interference based pressure model. Applied Acoustics 44, 53-65, 1995.
[15] S M Dance and B M Shield. Computer prediction of sound propagation in enclosed spaces using a pressure based model. Proc. Fifth International Congress on Air and Structure-borne Sound and Vibration, Montreal, 765-772, 1997.
[16] S M Dance and B M Shield. Modelling of sound fields in enclosed spaces with absorbent room surfaces. Part 1: Performance spaces. Applied Acoustics 58(1), 1-18, 1999
[17] S M Dance and B M Shield. Absorption Coefficients of Common Construction Materials for Use in Computer Modelling of Enclosed Spaces. Submitted to J. Building Acoustics, 1999
[18] J. Kang. Acoustics in long enclosures. J. Acoustical Society of America 99(2), 985-989, 1996.
[19] L. N. Yang, T. J. Cox and B. M. Shield. Modelling of sound propagation in underground stations. Proc. Inter-noise 96, 3173-3175, 1996.
[20] L N Yang and B M Shield. Development of a ray tracing computer model for the prediction of the sound field in long enclosures. J. Sound and Vibration, 1999 (in press)
[21] B M Shield and L N Yang. Modelling of acoustic parameters and speech intelligibility in long enclosures. Proc. 16th International Congress on Acoustics, 2157-2158, 1998.
[22] B M Shield and L N Yang. Two methods of modelling a curved surface in a ray tracing model. J. Building Acoustics 5(2), 69-78, 1998.

S682/008/99 Technical Note

Aircraft noise

A J KEMPTON
Rolls-Royce plc, Derby, UK

1 SYNOPSIS

This paper examines progress in jet aircraft noise reduction over the past thirty years, and describes current research in the UK and Europe aimed at further noise reductions. Research described is aimed at turbomachinery noise (through improvements in component design and attenuation as the sound propagates along the ducts) and reducing jet noise.

2 AIRCRAFT NOISE

Both the engine and airframe designs are important in determining the total aircraft noise. The elements of the engine that generate the most noise are the fan, the turbine and the jet, although compressor and combustor noise also contribute.

The advances in engine noise technology since the introduction of jet aircraft are illustrated by comparing the certificated noise levels over time for the full power sideline case, and normalising the results to a common aircraft thrust (see Fig. 1). A reduction of about 20dB in noise has been achieved since the early jet aircraft. The reductions in aircraft noise have been achieved both by increases in the engine bypass ratio and by advances in component designs.

Compared with early pure jets and low bypass-ratio engines, modern high and very-high bypass-ratio engines have significantly reduced jet velocities for the same thrust and consequently make much less jet noise (see Fig. 2). The reduction in jet noise due to increased bypass ratio, combined with advances in component design, results in a considerable reduction in total aircraft noise.

Forced mixers have been developed to reduce noise at take-off by up to 5dB (see Fig. 3). Turbomachinery noise is controlled at source by choice of blade numbers (Fig. 4), and as it propagates along the ducts by acoustic treatment along the duct walls (Fig. 5).

Rolls-Royce's objective is to develop the technology by 2010 to reduce aircraft noise by a further 10dB. To achieve this, it is pursuing a long-term strategy with an ongoing research programme to deliver noise reductions in the shorter and longer term. Technologies under consideration include the swept fan (Fig. 6), negatively-scarfed intakes (Fig. 7) and serrated nozzle lip geometries. It is also playing a major role in the first £20m phase of a European collaborative research programme into aircraft noise. This research is aimed at both the shorter term (demonstrating by 2005 the technology to deliver aircraft noise reductions of 6dB at both take-off and landing), and the longer term, assessing concepts such as active control and novel integrated engine/aircraft configurations.

3 CONCLUSION

Significant progress has been made in reducing aircraft noise over the past thirty years, and Rolls-Royce is investing in extensive research programmes to deliver major further improvements.

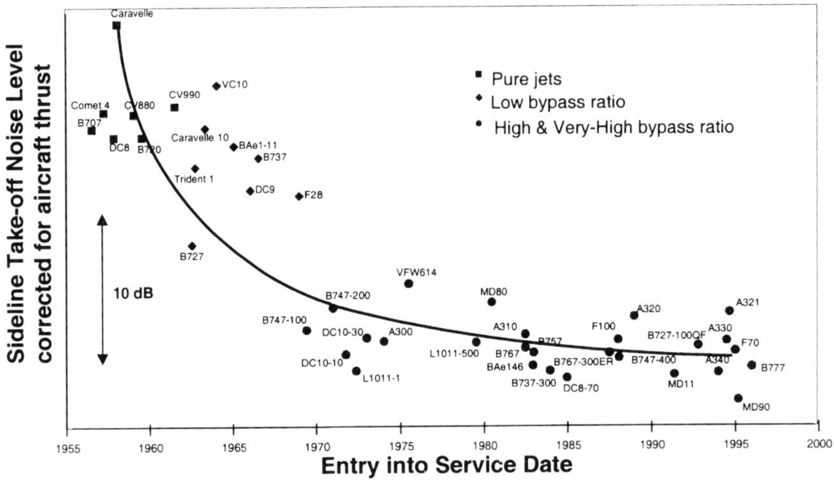

Figure 1: Corrected noise levels versus time

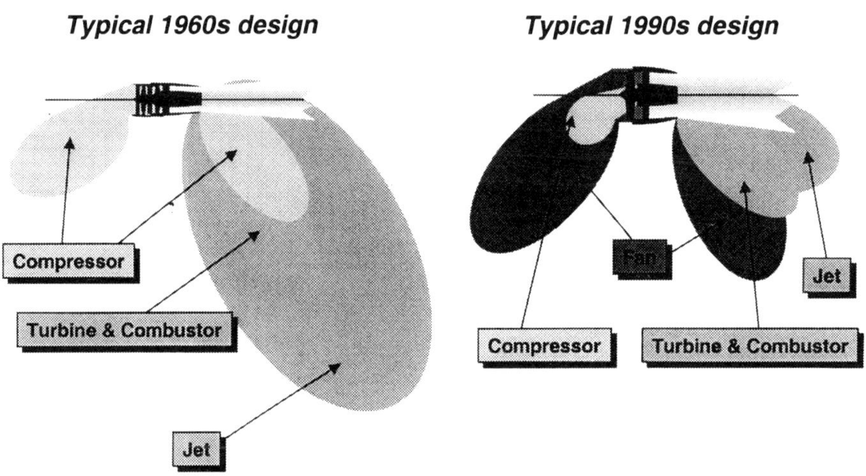

Figure 2: Main sources of engine noise

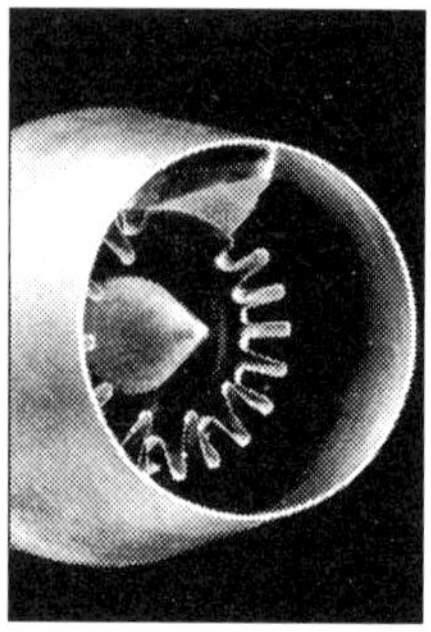

Example of a forced mixer.

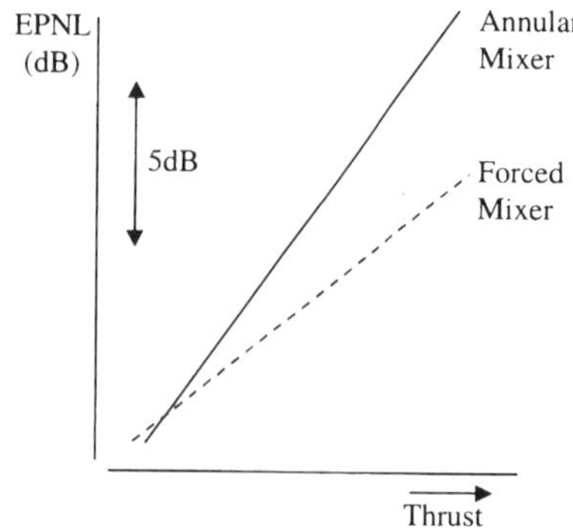

Noise benefits of forced mixer

Figure 3: Jet noise reduction with forced mixers

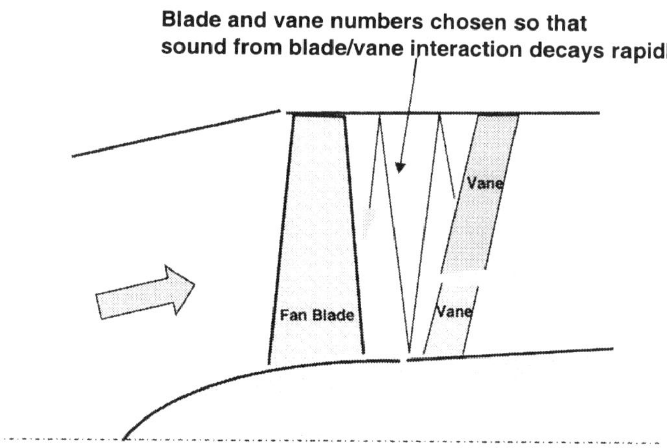

Figure 4: Turbomachinery noise reduction by choice of blade numbers

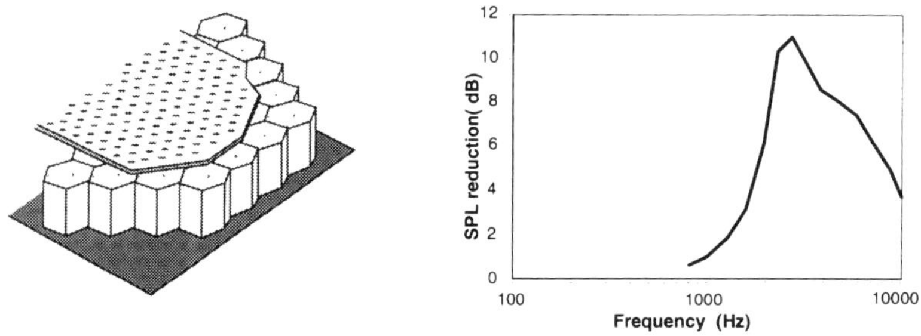

Typical acoustic liner

Typical attenuation spectrum

Figure 5: Turbomachinery noise reduction by acoustic treatment on duct walls

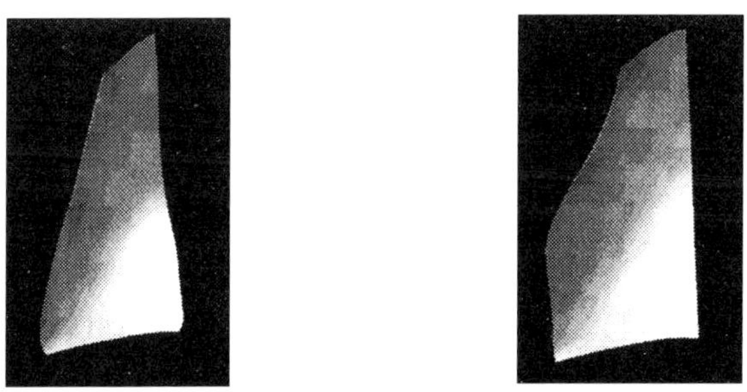

Conventional Fan **Swept Fan**

Figure 6: Swept fan design for low noise

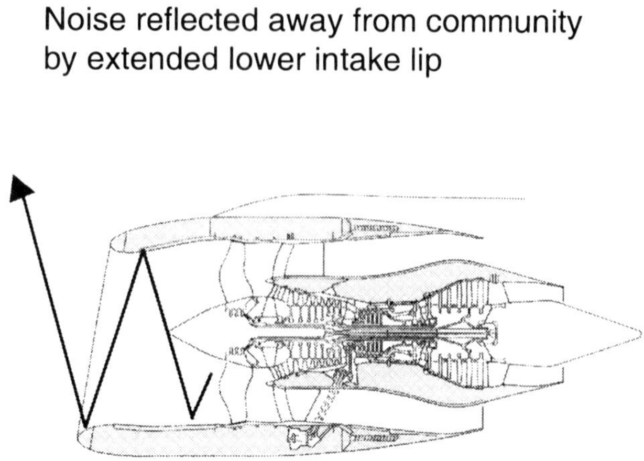

Figure 7: Negatively scarfed intake for low noise

Authors' Index

A

Attenborough, K 29–38

C

Cremers, L... 1–14

D

Dance, S 77–84
Davies, P O A L.............................. 39–50

G

Guisset, P 1–14

H

Hemsworth, B 69–76
Holland, K R 39–50

K

Kempton, A J 85–90
Krylov, V 51–68

M

McCulloch, C F 1–14, 15–28

S

Shield, B M...................................... 77–84

V

van der Walt, D 39–50
van der Linden, P............................. 15–28

W

Wyckaert, K...................................... 15–28

Y

Yang, L N 77–84

Organizing Committee Member

Eur Ing V A Cole
HM Principal Specialist Inspector of Health & Safety
(Mechanical Engineering)
Basingstoke, UK

Based on papers presented at the one-day Seminar *Noise and Vibration – Advances in Research and Development*, held at the IMechE Headquarters, London, UK, on 15 September 1999.